RHS

50 WAYS to OUTSMART a SQUIRREL and other GARDEN PESTS

Ingenious ways to protect your garden without harming wildlife

Simon Akeroyd

MITCHELL BEAZLEY

Contents

CHAPTER FOUR

HOW TO STOP SLUGS AND OTHER MEDDLESOME MOLLUSCS

CHAPTER FIVE

HOW TO OUTWIT ANTS AND OTHER INVASIVE INVERTEBRATES

Introduction

ONE OF THE MANY JOYS OF OWNING A GARDEN IS OBSERVING AND SHARING YOUR OUTSIDE SPACE WITH WILDLIFE. HOWEVER, IT IS NOT ALWAYS A MUTUALLY HAPPY RELATIONSHIP, WITH SOME UNINVITED VISITORS GETTING TO ENJOY A LARGER SLICE OF THE HORTICULTURAL PIE THAN YOU DO YOURSELF. THERE IS NOTHING MORE FRUSTRATING THAN SPENDING MONTHS NURTURING A PLANT ONLY TO FIND IT DESTROYED THE MOMENT YOU TURN YOUR BACK.

These unwanted assaults on your precious plants can come from creatures of all sizes, from daring deer to microscopic mites. To make things harder, attacks can come from all angles, with some of the pests flying overhead and others tunnelling underground. Some can scale vertical surfaces to devour your dahlias, while others will squeeze through the tiniest mesh for a munch on your marrows. Some attackers are slow and insidious and damage can go unnoticed for months, until there is enough accumulation for an infestation to show the harm done. Others are opportunists that can literally be passing through, yet within an hour or two have wreaked havoc in your garden and destroyed all of your plants.

Although many of the solutions are about protecting plants and crops from damage, other issues are addressed that aren't specifically plant based, such as avoiding wasps when dining al fresco, protecting birdseed from squirrels and dealing with anthills on the lawn. The solutions are practical and easy, and have been categorized as either a quick fix, fairly simple, or effort required, with nothing being too tricky or complicated. Most importantly, the chosen solutions are wildlife friendly and free from cruelty. The suggestions do not require any specialist garden or DIY expertise and should be easy to implement by following the simple step-by-step guides or the beautifully illustrated diagrams.

Once you have read this book, you should have the knowledge, skills and motivation to ensure you are never again outsmarted by a squirrel, or any other unwanted garden visitor.

Thankfully, there is help at hand in the shape of this book. Here, we address 50 common problems caused by wildlife or pets in the garden. The theory is that by getting to know your enemy, you can exploit their weaknesses and thereby protect your garden. Each solution explains the life cycles, habits and 'creature comforts' of many of the unwanted visitors to the garden or allotment, and provides practical and simple solutions to thwart their attempts at getting to enjoy your plants or outside space at your expense. The book is divided up into five easy-to-follow chapters based on the common attributes of the pest, such as their size, ability to fly or even their tunnelling skills.

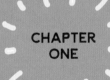

HOW TO COUNTER CATS AND OTHER UNWANTED VISITORS

Numerous creatures can create
havoc in your garden including
badgers, cats, deer, dogs, foxes and
moles. Discover how to prevent
them enjoying the garden
at your expense.

1 Defend your trees from deer damage

✳✳✳ effort required   🦌 deer

Deer are graceful creatures, and you should consider yourself lucky if they visit your garden. But you will need to take some action if you are to prevent them from destroying your plants.

OH DEER, WHAT IS THE DAMAGE?

Deer tend to graze on emerging shrubby shoots, quickly reducing young sapling trees and shrubs to jagged stumps. They are particularly fond of the young woody growth rich in sugary sap, which emerges in spring.

These mammals will dine on most plants – even thorny shrubs such as roses (*Rosa*) and blackberries (*Rubus*) or poisonous plants such as yew (*Taxus baccata*) are not safe from them. They will also happily munch through your veggies, fruit crop and herbaceous plants.

Types of deer

The deer most commonly found in the garden is the roe (*Capreolus capreolus*, pictured) or the smaller muntjac (*Muntiacus*). Occasional visitors are the timid fallow (*Dama dama*) and sika (*Curvus nippon*). The much larger red deer (*Curvus elaphus*) is commonly seen in Scotland and prefers the countryside to gardens.

Put up a fence to protect from deer

Erecting a fence is the most effective way of protecting your plants from deer. Despite the initial outlay, a fence around your entire garden is money well spent, as it will save your plants from becoming munched. The fence will need to be fairly high, as deer are the tallest wildlife visitors to the garden, with some being able to jump up to 1.8m (5ft 10in).

YOU WILL NEED:
- TAPE MEASURE
- SPADE (OR AUGER)
- 2.5M (8FT) TREATED ROUND VERTICAL FENCE POSTS
- SPIRIT LEVEL
- POST RAMMER
- WIRE MESH FENCING
- HAMMER
- GALVANIZED STAPLES OR NAILS
- PEGS

Mark out with a tape measure where the fence posts will go. With a spade (or auger), create preliminary holes in these positions, then put the posts in place. Use a spirit level to check posts are level, and then firm the soil back around the base. Use a post rammer to drive the posts into the ground, leaving 2m (6$^{1}/_{2}$ft) above ground.

Stretch the wire mesh fencing out between the posts and use the hammer to secure galvanized staples or nails. The mesh size of the fencing ought to be 20 x 15cm (8 x 6in), although muntjac may need even smaller mesh (75mm; 3in). Avoid using barbed or horizontal wire with gaps of more than 30cm (12in) between each level, as deer can squeeze through them.

Muntjac can dig, so netting should be firmly pegged down every 40cm (16in) along the perimeter.

Observation

A successful fence should stop deer visiting your garden altogether. Whether that is a blessing or a disappointment will depend on your point of view.

PROTECT YOUR ASSETS

If you have a large garden, it may be too costly to fence the entire space. Instead, you could fence the most valuable individual tree specimens. To do this, you will need to place four posts around the tree in each corner to create a square at a height of 1.5m (4ft 11in). Use a small mesh (less than 75mm; 3in), so deer can't push their heads through the gaps.

HEDGE YOUR BETS

Alternatively, you could plant a hedge – it can be an effective barrier if grown densely, and is tall and compact enough to stop deer pushing through. You will need to purchase large mature hedges, unless you are willing to wait for them to grow. Hedges look more pleasant than wire fencing and are better for wildlife, but individual plants will need to be positioned closely together to ensure deer don't push through any gaps. Keep the tops of hedges regularly pruned to ensure the plants don't get too leggy at the base.

OTHER DETERRENTS

Get a dog Deer are very wary of dogs and are unlikely to enter the garden if they suspect one may be present. Ensure your dog can't escape from the garden to chase and attack any deer that may be around the perimeter of your garden.

Lights and sounds Deer are very timid creatures, and it is possible to install sensors that emit an ultrasonic sound if triggered, which is inaudible to humans. However, they may start to become familiar with the noise, making it eventually ineffective. The same can be said for flashing lights used as deer scarers. These can also annoy your neighbours if not positioned sensitively in the garden.

Scare with hair Some gardeners believe wrapping human hair in muslin bags or an old pair of nylon tights and leaving them at entrance points to the garden will deter deer, but these methods seem to have varying degrees of success.

Deer-resistant plants

Deer will feed on most garden plants but are less likely to touch those listed below. There are variables, such as time of year, proximity to the house and whether or not there are more tasty plants available.

Deer are less likely to dine on:

Barberry
(*Berberis vulgaris*)

SHRUBS, CLIMBERS AND TREES:
- Bachelor's buttons (*Kerria*)
- Barberry (*Berberis vulgaris*)
- Bay (*Laurus nobilis*)
- Butterfly bush (*Buddleja davidii*)
- Climbing honeysuckle (*Lonicera*)
- Cotoneaster (*Cotoneaster*)
- Cypress tree (*Cupressus*)
- Easter tree (*Forsythia*)
- Lavender (*Lavandula*)
- Mexican orange blossom (*Choisya*)
- Oregon grape (*Mahonia*)
- Pieris (*Pieris*)
- Rosemary (*Salvia rosmarinus*)
- St John's wort (*Hypericum*)
- Winter jasmine (*Jasminum nudiflorum*)

HERBACEOUS PERENNIALS:
- Blue bugle (*Ajuga*)
- Catmint (*Nepeta*)
- Christmas rose (*Helleborus*)
- Common peony (*Paeonia*)
- Foxglove (*Digitalis*)
- Lady's mantle (*Alchemilla*)
- Mexican fleabane (*Erigeron*)
- Monkshood (*Aconitum*)
- Red hot poker (*Kniphofia*)
- Spurge (*Euphorbia*)

HERBS, FRUIT AND VEGETABLES:
- Gooseberry (*Ribes uva-crispa*)
- Mint (*Mentha*)
- Oregano (*Origanum vulgare*)
- Rhubarb (*Rheum rhabarbarum*)
- Sage (*Salvia officinalis*)
- Thyme (*Thymus vulgaris*)

Bay
(*Laurus nobilis*)

2 Erect an electric fence to keep badgers at bay

✳ ✳ ✳ effort required   🦡 badgers

Have you ever had that sinking feeling when you realize your sweetcorn crop has been devoured by badgers? The most effective way to protect your fruit and veg is to put up an electric fence.

PAWS FOR THOUGHT

These stocky mammals are mainly nocturnal, so you are most likely to find them snuffling around the garden at night. They can dig up lawns, causing much destruction as they search for earthworms and leatherjackets. When these are difficult to find, due to frosts or dry weather, badgers can eat a wide variety of foods, including your flower bulbs, fruits and vegetables! They also dig repugnant latrines – yet another reason to want them off your plot.

A SHOCK TO THE SYSTEM

Currently, an electric fence is the most effective and humane deterrent. Electric fence materials vary according to the manufacturer, so always check instructions. Some come in kit form, while others require you to buy individual items.

Avoid touching the fence once it is live and attach a sign to warn others. An electric fence voltmeter can be used to ensure it is working. Always use battery energizers, as mains ones can give a very severe but harmless jolt. Keep vegetation cut back around the base, as this can cause the fence to stop working.

Setting up your fence

Many agricultural or hardware stores will supply kits to put up temporary electric fences. Always follow their instructions, as equipment will vary between different brands.

YOU WILL NEED:
- WOODEN SUPPORT POSTS
- ELECTRIC FENCE INSULATORS
- METAL WIRE
- ELECTRIC FENCE CHARGER
- TIMER SYSTEM (OPTIONAL)

Place wooden support posts around the area of garden you wish to protect at a spacing of about 2m (6^1/$_2$ft). Screw electric fence insulators into the posts starting at 10cm (4in) above ground, and then at 20cm (8in) intervals. Run a series of three parallel wires through these.

Connect the wires to the charger and then plug the charger into an appropriate power source as recommended by the manufacturer's instructions. Fix a timer to the plug if you wish to control the time when the electric fence is active.

ALTERNATIVE METHODS

If an electric fence is not suitable for your garden, there are other things to try.

Fenced in If you don't have an entrance to the badgers' sett in your garden, you could erect a sturdy garden fence to make it harder for them to get in. If you have a gate, make sure it is closed at night. For high-risk crops, use enclosures of chain link fencing or corrugated iron for vegetable and strawberry plots; wire netting is generally easily torn by badgers.

Deter with devices Try using ultrasonic devices that emit a high-pitched sound (inaudible to our ears), which scare the badgers when they get close. You can also try something similar with motion-triggered LED lights.

Spray away Some people swear by human urine. Male human urine watered down and sprayed around an important crop may be enough to keep territorial badgers away.

GET A WHIFF OF THIS

Badgers have a great sense of smell and use this for communication within and between social groups. They produce scents to act as warning signals, while others communicate things like mating status.

Hands off!

Badgers are protected by law, so you're not able to harm them or damage or obstruct a sett. You can, however, put them off key, high-value areas of your garden or allotment.

CLAN OF THE BADGERS

The Eurasian badger (*Meles meles*) is a member of the *Mustelid* family,
related to animals like otters and stoats. Badgers live in social groups called
clans of four to eight individuals on average, in an underground series of
chambers and tunnels called a sett. Larger setts can extend 20–100m
(65–328ft) in length! These homes can be passed down from generation
to generation of badger.

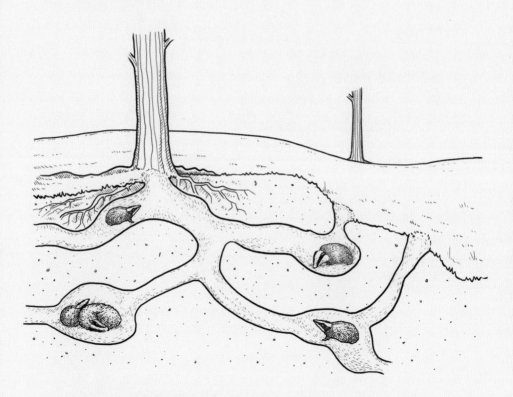

TIMING IS EVERYTHING

Your garden is unlikely to be visited by badgers over winter. In autumn they
will fatten up before a period of inactivity or torpor during the cold weather.
In February the females (sows) will usually give birth to two or three cubs,
which will remain in the sett for 12 weeks, before braving the big, wide world.

3 Stop foxes digging up your garden

** Fairly simple   🦊 Foxes

For many people, watching foxes playing in the garden is a pleasure, but the downside is that they can cause damage. To outsmart them, you have to be as cunning as, well, a fox…

TELL-TALE SIGNS

Despite their cute faces and bushy tails, foxes can be a menace in the garden, particularly if they create a den nearby and have a litter of noisy, playful cubs. As well as digging up plants, other annoyances include leaving excrement about the place and burying 'stolen' items such as food, bones, rabbits, shoes, gardening hand tools or shiny toys. They also cause damage with their teeth, by chewing on hosepipes for example.

Holes dug by foxes in the lawn look messy and could cause you to twist your ankle if you are not watching your step. If you are a light sleeper, you may find their yelping noises keep you awake at night. Male foxes may also mark their territory with urine, which has an extremely pungent aroma.

WHAT TO DO?

As a rule of thumb, foxes have to be tolerated if they regularly visit a garden. It is practically impossible to erect barriers, as they both dig under or climb over fences. However, there are a few things that can be done to mitigate their negative impact and damage.

Did you know?

While most members of the dog family, such as wolves, jackals, coyotes and dingoes, hunt as a pack, the urban fox is predominately solitary and generally forages alone.

Outfoxing these cunning canines

- **CHAFER GRUB CRAVINGS**
  Foxes love to feast on chafer grubs, and they will tear up your lawn in order to satisfy their appetite for them. If you are able to control chaffer infestations in the lawn, you will also limit fox damage. Using a natural predator such as nematodes (see page 96), which can be applied to the lawn in late summer, will help manage chafer numbers.

- **FERTILIZER NO-NO**
  Avoid using fertilizers made from animal products such as bonemeal, chicken pellets or blood, fish and bone, etc., as the aroma attracts foxes, encouraging them to dig up plants in the hope of finding food.

- **GET CAGEY**
  Build a cage or wooden structure out of pallets to protect your bins. Rubbish is a main attraction for foxes to your garden – they will rip bin bags apart if left out. If you have a wheelie bin, place a weight on top of the lid, as foxes have been known to nudge open and flip back the lid to access the contents. Don't place wheelie bins next to fences or walls, as foxes use these to climb on, lean over and open the lids.

- **SCENT AND SOUND**
  There are many animal-deterrent products on the market. Although it is unclear if they are effective, they will do no harm. You will need to regularly reapply scent products to prevent them from wearing off. There are also products that emit an ultrasonic sound, which humans can't hear but foxes can. This can be less effective as foxes become familiar with the noise.

← Foxes will always be able to get into your garden, no matter what type of fence you erect to try to keep them out.

4 Keep plants out of the reach of rabbits

✳✳✳ effort required 🐇 rabbits

Creating a raised bed to keep hungry rabbits off your veggies is both an attractive and practical solution.

REACHING NEW HEIGHTS

Not only can raised beds of around 1m (3ft 3in) make it harder for rabbits to reach and devour your plants, but there are other advantages to having them in your garden. First, they can be made at a comfortable height for you to tend to plants, helping to avoid backache. Second, they provide decent drainage as the soil warms up faster. They also make attractive features and can be used as focal points or create a structure to your overall garden design.

Raised beds can be made from most materials, but many are constructed from exterior timber or bricks. You can also simply construct a raised bed from rocks in the shape of a keyhole, hence the name: keyhole gardening.

SHAPING YOUR GARDEN

Keyhole gardening was developed in Africa, where people constructed raised beds using materials found in the landscape. Without using this method, the existing soil lacks nutrients and structure, making it unsuitable for growing food in.

The main feature of a keyhole raised bed is a compost heap in the centre, where nutrients from the decomposing material leach out into the soil, providing fertile growing conditions.

Did you know?

One of the reasons rabbits are constantly gnawing away at abrasive material such as tree bark is because their teeth never stop growing – they can grow at a rate of 12cm (5in) a year. If they were not constantly chewing, their teeth would grow so large that they would not be able to eat.

How to create a keyhole garden

YOU WILL NEED:
- A STICK
- WOVEN GRASS (OR CHICKEN WIRE)
- THREE 1M (3FT 3IN) POSTS
- ROCKS AND STONES OF VARYING SIZES
- MORTAR (OPTIONAL)
- SOIL OR COMPOST
- MATERIAL FOR THE COMPOST HEAP

Mark out the shape of a keyhole in the ground using a stick. An ideal diameter for a keyhole bed is about 3m (10ft) across. Construct the compost heap in the centre and make it from permeable material so that nutrients can leach out into the soil. You can use woven grass, but chicken wire is just as good. Knock the three posts into the ground to create a circle with a 75cm (30in) diameter. Attach chicken wire around the outside of the posts to contain the material for composting.

Create the circular wall of your keyhole raised bed using rocks and stones. Leave a 'walkway' to access the compost in the centre. Aim for a height of around 1m (3ft 3in). Use the largest rocks on the bottom and smaller ones towards the top. You shouldn't need mortar, but you could use it to hold the rocks together. Fill the raised bed with soil or compost to the height of the walls. Allow the soil to settle before starting to plant the bed up. Fill the compost heap with material that will eventually start to decompose, leach into the soil and feed the plants.

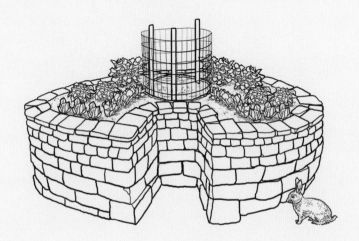

5 Protect your plants from being nibbled

✳︎✳︎✳︎ effort required   🐰 rabbits

The last thing you want is for uninvited guests to destroy all your hard work in the garden. But there is a way to keep your plants safe from hungry rabbits without harming them.

AN APPETITE FOR DESTRUCTION

Rabbits might look cute, but they can wreak havoc in the garden. They chew on woody stems and tree trunks, and they also dig up herbaceous plants. Rabbits love nothing better than grazing on vegetables or young emerging shoots, and will even dig holes in the lawn.

These little mammals are at their worst in spring or summer when the first of their young have been born. However, they are more likely to damage the base of trees and shrubs during winter when they have no herbaceous material to nibble on. Once a rabbit has chewed around a trunk (known as girdling), the plant will die.

Did you know?

Rabbits breed like, well, rabbits! It is estimated that a mating pair can have up to 65 babies in one season. Considering all female babies (called kittens or kits) are also capable of breeding within three months, they could soon be running riot in your garden.

BUILD A BARRICADE

The most effective way of keeping your plants safe from bunny attacks is to erect a wire mesh fence around the edge of the garden. In a large garden, it might be worth placing tree guards around individual specimens instead.

Before constructing a fence, make sure that the rabbits' burrow is not situated inside the garden; otherwise, all you will be doing is fencing them in so they are free to munch away. See overleaf on how to build a fence.

Build a fence to keep rabbits out

YOU WILL NEED:
- WIRE FENCING WITH 2.5CM (1IN) MESH
- STAKES
- HIGH-TENSILE STRAINING WIRE
- STAPLE GUN
- LARGE HAMMER
- SPADE

The rabbit fence needs to be about 1m (3ft 3in) above ground height. Mesh comes in different sizes; don't use traditional chicken wire, as a rabbit can slip through the gaps. Instead, choose a mesh that has smaller gaps.

Rabbits love to burrow, so the fence needs to be dug 30cm (12in) underground, to discourage them from burrowing underneath. Curve the bottom of the underground section 15cm (6in) outwards, away from the fence. To keep the fence upright, staple the mesh to solid stakes embedded into the ground at 2.5m (8ft) intervals. Run a taut wire around the top of the mesh and fasten them together.

Regularly check around the perimeter of rabbit fences, as rabbits are persistent creatures that will continually scratch and dig at the surface to exploit any weak mesh. Take care too in winter, as they can use snow to overcome the barrier.

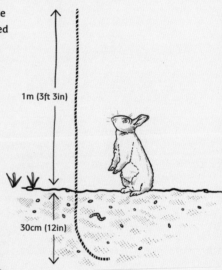

1m (3ft 3in)

30cm (12in)

Don't forget

Remember to add a gate to your fence – and make sure that this, too, is rabbit-proof.

6 Keep cats out of your back yard

✳ ✳ fairly simple   🐈 cats

If cats are making a nuisance of themselves in your garden, there are a number of innovative methods you can try to deter them, including creating a thorny environment and using citrus peel.

FELINE TRESPASSERS

If cats are treating your garden like a public lavatory, there are a few techniques worth trying to put them off. As cats are carnivores, their faeces carry pathogens and parasites, and can contribute to the spread of toxoplasmosis – something you certainly don't want near any of the food crops you intend to eat.

THE BIG HOSE DOWN

Cats are creatures of habit, so they scent their territory regularly when they visit, and frequent the same area time and time again. If you hose down the area regularly, you will wash away their aroma, and they will tend to be less drawn to return next time. See overleaf for more cat deterrents.

Did you know?

The scaredy cat plant (*Plectranthus caninus*) smells of dog urine and discourages both cats and dogs. Its pungent aroma can also be smelt by humans, so it is perhaps not a plant to grow near your patio. You will need to overwinter (keep it alive through the winter) it indoors or buy it afresh each summer.

Hawthorn
(*Crataegus*)

Holly
(*Ilex*)

HOW TO STOP FELINE VISITORS

Get digging Our feline friends love nothing more than freshly dug-over soil to go to the toilet in. So, if you create a small patch of about 30 x 30cm (12 x 12in), away from your other plants, the chances are the cat will go there and leave your other flower beds alone. Lightly dig it over and rake it level.

Thorny relationship Cats like to get comfy when going to the toilet, so if you make it as uncomfortable for them as possible, they are likely to go elsewhere. Try cutting lengths of prickly shrubs such as barberry (*Berberis vulgaris*), brambles (*Rubus fruticosus*), gooseberry (*Ribes uva-crispa*), hawthorn (*Crataegus*), holly (*Ilex*) or roses (*Rosa*), or anything else with sharp thorns; lay them on top of areas you have recently dug over or sown seeds into. Chicken wire laid across the beds is also uncomfortable for cats, and plants will usually grow through it too.

Get some gadgets Motion-activated water sprays can help to keep cats at bay, while motion-triggered gadgets that emit an ultrasonic sound, inaudible to humans but unpleasant to cats, can also deter them.

Lemons and limes For some reason, cats hate the smell of citrus fruit. Therefore, if you have any orange (*Citrus × sinensis*), lemon (*Citrus × limon*) or lime (*Citrus × aurantiifolia*) peels, place them on flower beds at 10cm (4in) intervals to put cats off visiting. If you don't have any peelings, then try using a few drops of lemon juice from a squeezy bottle.

Heaven scent

Cats are likely to steer well clear of the following things because they don't like their smell:

- Citronella (*Pelargonium 'Citronella'*)

- Lavender (*Lavandula*)

- Lemon thyme (*Thymus citriodorus*)

- Pennyroyal (*Mentha pulegium*)

Alternatively, you could try human hair or used coffee grounds, or buy commercial feline repellents that mimic the scent of a predator, for example, by containing lion dung.

Lavender
(*Lavandula*)

7 Grow catnip to entice cats away from your plants

✳ quick fix   🐈 cats

Lure feline friends away from your vegetable patch by planting a sacrificial bed full of catnip, which cats adore.

EAGER EXPLORERS

Whether it is your own cat or your neighbour's, it is practically impossible to keep them out of your garden due to their agility to climb trees and over fences. While you can erect fruit cages around specific flower beds to stop them fouling among your plants, it isn't practical to cage your entire garden – and it wouldn't look good, even if you were able to. Furthermore, fruit cages are expensive, even for the smallest of flower beds.

A SMALL SACRIFICE

There is an old traditional remedy that can help – it is easy to implement and fairly cheap. It involves planting catnip, which members of the feline family love. The theory is that the cats will enter your garden and be so drawn to it that they will keep away from the other areas.

Did you know?

The compound nepetalactone, found in catnip, also repels insects such as aphids, so it's worth planting somewhere in your garden where aphids may be a problem, such as on broad beans (*Vicia faba*) or roses (*Rosa*).

Catnip
(*Nepeta cataria*)

Catmint
(*Nepeta mussinii*)

CATNIP OR CATMINT?

There are two similar plants from the mint family, and both are from the genus *Nepeta*; one is commonly known as catnip, the other catmint.

Catnip is *Nepeta cataria*, a short-lived herbaceous perennial, and this is the one that cats crave, making them euphoric. It contains a feline-attractant compound called nepetalactone, which gives cats a natural 'drug'-induced high.

Catmint is *Nepeta mussinii*, and although cats seem to be very fond of it – to the extent that they will roll around in the plants – it does not seem to have the euphoric effect that catnip has.

PLANT PLACEMENT

Both catnip and catmint prefer well-drained soil in dry, sunny locations. Catnip will produce white flowers with light purple spots and has a slightly 'weedy' appearance. Catmint, on the other hand, has attractive lavender-coloured flowers and has a lax habit, perfect for edging a path or at the front of a border. It also attracts bees.

So, as a rule of thumb, if you want a more attractive, ornamental plant, choose catmint; if you want something to really entice your cat away from your veggie plants, choose catnip.

Plant catnip as far away as possible from the plants you are trying to protect. The catnip should detract and divert cats from other plants.

8 Weave a fence to keep dogs out

✳ quick fix dogs

Construct a beautiful woven fence in your garden to stop your dog running over flower beds – it also makes for an eye-catching, rustic feature.

WEAVE YOUR MAGIC

Low, woven fences are a wonderful way of giving definition to a flower bed and creating a bit of height and interest to what might otherwise be just a flat space. Woven fences are also very flexible in that you can easily make them follow the contours of curvy flower beds. If you have enough material, you can make them to any height simply by adding more flexible stems.

Willow fences are partly permeable, meaning they make good wind breaks, as they slow the wind down but still allow air to circulate, which is important for reducing the effect of fungal diseases and some pests.

A living hedge

Make a living willow hedge by pushing the base of young withies (strong, flexible stems) about 10cm (4in) into the ground at 15cm (6in) intervals. In spring they will produce foliage and new shoots, which can be crossed and tied to make a pattern or woven together. Don't plant it close to other plants as willows suck up a lot of moisture from the ground. Willow grows tall and fast, so may need some heavy cutting back.

Making your fence

YOU WILL NEED:
- LUMP HAMMER
- SOLID UPRIGHT STAKES
- PLANTS FOR WEAVING (SEE BOX, RIGHT)
- SECATEURS, LOPPERS OR SAW

Use a lump hammer to knock the upright stakes into the ground at intervals of about 40cm (16in) along where you wish to construct your fence. These can be straight stems of hazel (older than those used for weaving) with at least 2.5cm (1in) diameter, or uprights can be bought from garden centres. Sweet chestnut (*Castanea sativa*) stakes are a good alternative. Their height depends on how high you want the fence – about a quarter of its height will need to be pushed into the ground.

Weave the lengths of willow or hazel through the stakes; alternate the side each length is started to create a stronger structure. Continue adding layers of stems to the top of the fence until you have reached the desired height. Use secateurs or loppers to trim off the ends to provide a neat finish. Use a saw or loppers to cut the tops of the upright stakes so they are flush with the top level of the woven stems.

Plant preference

Traditionally, hazel (*Corylus*) or willow (*Salix*) stems (known as 'withies') are the best plants to use for weaving. These should be one or two years old as they are the most flexible. Cut stems in late winter just as the sap is about to rise, meaning they are nice and supple. If you are not using them straight away, place in water for a few days so they retain their pliability.

Young, colourful dogwood (*Cornus*) can also be used. It provides a vibrant fence in a range of hues, although the colours fade after a few months.

If you have a willow or hazel bush in your garden, you could cut your own stems. Alternatively, you can buy cheap bundles of young stems in winter from garden stores or online.

9 Stop dog urine damaging your lawn

✳✳✳ effort required dogs

People adore their pet dogs, but one downside of having a female is that they urinate on the lawn, causing unsightly brown burn patches. Combining tricks to prevent this with laying new turf will ensure a garden with a verdant feel – and no brown marks.

Replenish your lawn

To restore your lawn to its former glory, you will need to dig out the dead, brown, burnt urine patches and replace them with new turf.

YOU WILL NEED:
- SPADE OR HALF-MOON
- FRESH ROLL OF TURF
- RAKE

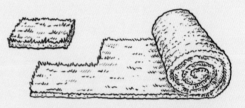

Use a spade to remove the dead section of turf in the lawn. Try to keep the removed patch in one piece and in a regular shape, such as a square. Using a fresh roll of turf, use the removed section of dead turf as a template, and cut around it with a spade or half-moon.

Rake the surface of the soil where the dead piece of turf has been removed to create a fine tilth and get it level. Place the new piece of turf over the hole and tamp it down with the back of the rake to firm it into place. Water the turf if no rain is expected.

Reap what you sow

Instead of buying turf, sow grass seed in the patch. Boxes of seed are much cheaper than rolls of turf and can be stored in the shed for months.

HOW GREEN IS YOUR GRASS?

Lawns come under a lot of criticism due to their lack of environmental credentials. To remain green and tidy, they usually require regular mowing, constant feeding with fertilizers, regular irrigation and use of weedkillers. Environmentally it is better to allow lawns to grow naturally, avoiding adding artificial substances and not using petrol-based machinery.

There are many positive green aspects to having grass. After all, every tiny square inch supports hundreds of grasses all pumping out oxygen into the air and absorbing carbon dioxide. A lawn also has a cooling effect on the atmosphere, reducing global warming and can minimize flooding as it absorbs moisture, reducing run-off. See overleaf for ideas on how to help your lawn.

Getting away from grass

You could consider a different type of garden – there are many other materials that your dog can't ruin:

DECKING
Natural-looking wood can add a rustic or chic style to your garden. Just ensure it is constructed from non-slip material, as decking can be slippery when wet.

PATIO
Stone, concrete or even terracotta slabs are hard wearing and require little maintenance.

BARK, CHIPPING OR GRAVEL
Wood bark or chipping provides a natural surface for shaded or woodland areas, while gravel is a great alternative for sunny locations.

Say goodbye to brown patches

Try these suggestions if dogs have urinated over your lawn and created unsightly burnt patches.

- **PROPRIETARY PRODUCTS**
  There are a number of products that can be bought from garden centres or online to help restore the lawn back to its verdant self. These can neutralize the nitrogen and salts from the dog's urine.

- **DOG'S DINNER**
  There are products that can be added to dog food or water to reduce the levels of nitrogen in their urine – consult with a vet before taking this course.

- **SPRAY AWAY**
  Motion-sensitive sprinklers that squirt a jet of water at your dog if they venture onto areas of the lawn can be effective.

- **LAWN FEED**
  Stop fertilizing with lawn feed, as the additional nitrogen from these feeds exacerbates the problem already caused by the nitrogen-rich urine.

- **MALE VS FEMALE**
  Be aware that female dogs tend to urinate in one spot, concentrating the problem. Males mark their territory and urinate in smaller amounts in different places.

- **DOG TRAINING**
  Train your dog to urinate in a designated spot.

Turn your garden into a meadow

If you let your grass grow a bit longer and sow some wildflowers,
it will attract a rich biodiversity of butterflies, bees and other pollinators.
You will no longer need to mow or feed it, and if your dog does decide
to urinate on the grass, it will not show up.

YOU WILL NEED:
- SPADE
- GARDEN FORK
- RAKE
- WILDFLOWER MIX
- FLEECE
- WIRE-MESH FENCE
- STRIMMER OR SCYTHE

Remove the top layer of the lawn with a spade. Fork over the exposed soil
down to a depth of around 10cm (4in) and remove any perennial weed roots.
Level it using a rake and break the surface down to a fine tilth. Consider
skimming off turf to lower the fertility as wildflowers prefer low-fertile soils.

Sow an annual wildflower mix into the soil at the rates recommended on the
packet. Until the meadow is established, you may need to place a fleece over
it to protect it from birds and give the seeds a boost. You could also erect
a temporary wire-mesh fence to prevent pet dogs running over it.

Enjoy the meadow in summer; in late summer, cut it down with a strimmer
or scythe. Leave the cut meadow on the surface for a couple of weeks before
removing. This gives seeds time to drop into the soil, which will provide you
with wildflowers for next year.

10 Use cunning to ward off meddlesome moles

✳✳✳ effort required   🦔 moles

This underground mammal is as smooth an operator as its soft, silky coat. To outsmart a mole, you must approach with patience, stealth and cunning.

GOING UNDERGROUND

Damage to plants in the garden by moles is accidental. Moles are carnivores, mainly feeding on worms and underground grubs – they don't actually eat plants. However, their tunnelling can destroy roots, causing plants to eventually die. They also create havoc on the surface of a lawn, creating large mounds of soil that they have excavated while creating an underground network of tunnels.

Molehills make it impossible to mow the lawn, they look unsightly and also cause the grass to die underneath the soil. Most tunnels are 35–75cm (14–30in) below ground, but shallow tunnels just below the surface can cause humans to twist their ankles on uneven and unstable ground, particularly if the lawn gives way.

FLYING SOLO

Moles are very territorial, and although the number of molehills can make it seem like there are a lot of them underground, there is usually only one mole in an average-sized garden. They are solitary creatures, and if they encounter another mole on their 'patch', they will often fight to the death. They usually only purposefully meet another mole for breeding, but apart from that live alone.

SEASONAL DAMAGE

Mole damage is usually at its worse in the winter because they feed on worms, which are frequently found in moist soil. During the summer, the soil in open ground is often dry, so moles disappear under hedges and shrubs, where their damage isn't so noticeable. However, well-watered vegetable gardens are at risk.

In winter, the damp soil in open ground is irresistible, and they start tunnelling in places such as the centre of the lawn. See overleaf to discover how to keep moles out of your garden.

Flexible fur

A mole's silky-smooth fur is very unusual, as it can be brushed both forwards and backwards. This is because moles go up and down tunnels, and their flexible fur avoids ruffling or 'backcombing' when going in reverse.

BLOCK THEM OUT

One of the most effective methods of keeping moles away is to place mesh netting underneath the lawn. This doesn't harm the mole; it simply encourages it to go elsewhere. Anti-mole netting is durable and rip-proof, and can be purchased online or from garden centres.

To install the netting, use a spade to remove the existing turf. On larger lawns, you can hire a turf-lifting machine. Dig down 10cm (4in) and place the soil in a heap to use later. Roll out your anti-mole net and peg it down. Overlap additional sections of netting by 10cm (4in), then add back 10cm of soil. You are then ready to re-turf or seed your lawn.

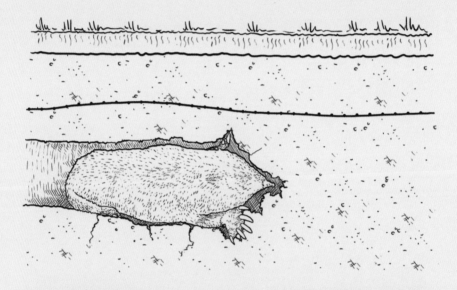

On a positive note

The excavated soil in a molehill has usually been broken up into a fine tilth and is perfect for using as part of a potting compost or as top dressing when levelling out dips and hollows in the lawn.

How to outsmart a mole

Here are some tricks you can try to send moles on their way:

- **PLANT CAPER SPURGE**
  It is believed that moles do not like the scent of this biennial root, so plant them where moles are a problem. Be aware that this plant can be invasive, the seedling could be more of a nuisance than the moles themselves, and it can also be harmful to people.

- **GROW YELLOW GARLIC**
  This is another plant whose aroma is said to be disliked by moles. Buy balls coated in its scent and bury them in the soil. Some people swear by boiling up garlic bulbs (to prevent them sprouting) and burying them underground.

- **BUY MOLE SCATTER GRANULES**
  Sprinkle these biodegradable and natural granules around molehills. Coated in natural plant oils, it has been suggested that the smell taints the soil and moles move away. Water the granules after sprinkling to ensure the smell percolates down into the tunnels.

Caper spurge
(*Euphorbia lathyris*)

Yellow garlic
(*Allium moly*)

- **ERECT SOLAR VIBRATING POLES**
  Insert solar vibrating poles into tunnels and molehills – it is claimed the vibrations and sound pulses scare the timid creatures away. Try one pole every 15m (50ft) radius.

CHAPTER TWO

HOW TO OUTSMART SQUIRRELS AND OTHER CRAFTY RODENTS

Squirrels and similarly shrewd
garden visitors, including rats and
mice, can get up to no end of
mischief. Learn how to outdo them
using the ultimate tricks in
cunning and deception.

11 Save your bulbs from squirrels

✳✳ fairly simple squirrels

Squirrels are typical hoarders and will happily dig up your freshly planted bulbs to stash away for later. However, there are a few simple things you can do to thwart their bulb-stealing tendencies.

THINK DEEP

Surprisingly, squirrels do not dig large holes – they will often only probe a few centimetres below the surface. Therefore, if squirrels are a problem in your garden, try planting bulbs slightly deeper than you usually would. It might take them longer to grow to the surface, but they will at least not get attacked by squirrels.

COVER UP

Squirrels seem to be attracted to freshly dug soil, detecting there will be bulbs underneath. Covering over freshly dug soil with leaf litter or compost can hide the fact that there is anything of interest underneath. Squirrels also dislike digging through the roots and shoots of other plants, so if possible, bury bulbs in among underground cover, or where there are nearby tree roots, taking care not to harm the tree.

Bulb choice

You could choose to use bulbs that squirrels do not usually like. Some examples are:

- *Alliums*
- *Hyacinths*
- *Muscari*
- *Narcissus*

Seal bulbs away

Keep your bulbs safe by sealing them in aquatic baskets underground.

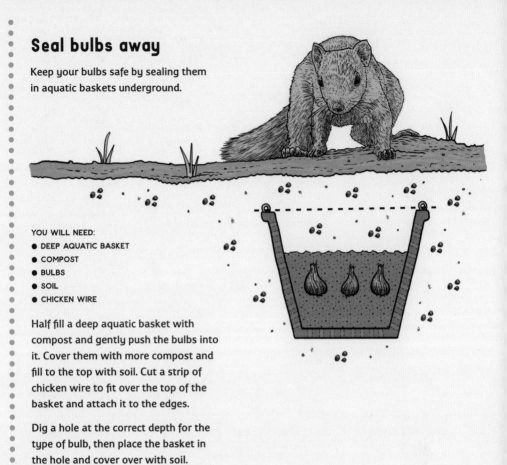

YOU WILL NEED:
- DEEP AQUATIC BASKET
- COMPOST
- BULBS
- SOIL
- CHICKEN WIRE

Half fill a deep aquatic basket with compost and gently push the bulbs into it. Cover them with more compost and fill to the top with soil. Cut a strip of chicken wire to fit over the top of the basket and attach it to the edges.

Dig a hole at the correct depth for the type of bulb, then place the basket in the hole and cover over with soil.

PROTECT WITH MESH

Another trick is to dig out an area for the bulbs, place them in the hole and then place a sheet of chicken wire just above them. This can be pegged down with tent pegs. Cover the chicken wire with soil. The bulbs will grow up through the mesh and the squirrels will not be able to dig them up. Alternatively, lay the mesh on top of the soil and carefully remove it once the bulbs start growing, as squirrels seem to lose interest when the bulbs are not dormant.

12 Protect nesting songbirds

*** effort required   squirrels (and other nest predators)

Would you like to encourage birds to nest in your garden by putting up a nest box, but worry about the safety of the chicks? Help your feathered friends with some of these top tips.

LURKING DANGERS

It's a tough job for birds to raise their young, for chicks to fledge the nest and survive in the great outdoors, but we can help them by putting up nest boxes. Unfortunately, and a danger to every nest are predators like squirrels, cats, woodpeckers, corvids (a bird of the crow family) and rats. Luckily, there are a few solutions to help.

QUID PRO QUO

By protecting your nesting songbirds, you will also help to control pesky invertebrates like aphids and caterpillars. Most songbirds will feed their chicks on these invertebrates, which provide a good source of protein for the growing youngsters. A blue tit, for example, usually lays between eight to twelve eggs. Once those eggs hatch, the young will need feeding, and it's thought that each chick can eat around 100 caterpillars a day!

Did you know?

To maximize the chance of a bird using your nest box, place it out of direct sunlight, where it will be sheltered from the weather – between a north and east aspect is best.

CHOOSE THE BEST BOX

Buy for the breed Buy a nest box to suit the species of bird you want to attract. If you want to entice the robin to your garden, for example, try the open-fronted shelf sort.

Size matters If you get a nest box with an entrance hole, it needs to be big enough for the nesting birds to get in, but small enough to keep other things out! The size will depend on which birds you want to attract:

- 25mm (1in) diameter for very small birds such as blue tits and coal tits.

- 28mm (1⅛in) diameter for small birds like great tits.

- 33mm (1⅓in) diameter for slightly bulkier birds like house sparrows and nuthatches.

Check the entrance hole When buying or making a nest box, ensure the entrance hole is positioned at least 120mm (4½in) up from the bottom. This thwarts nest raiders from easily reaching in.

Avoid a perch Nest boxes that come with a perch on the front do not help the nesting birds, but can make it easier for predators to get in, so buy one without.

Use metal plates These stop predators from enlarging the entrance hole to gain access.

Place with care Aim for 2–4m (6–13ft) from the ground on a tree or a wall, and try to keep the entrance hole clear from clutter. Feeding stations attract lots of birds and other wildlife, so bird boxes close by could suffer from disturbance and make it more noticeable to potential predators.

13 Turn up the heat on squirrels

✳ quick fix   🐿 squirrels

**Tired of watching greedy squirrels steal the bird food?
Try spicing things up by adding some chilli.**

DISRUPTING THE CALM

Sometimes there is nothing better than watching birds flit back and forth between the hedge and the feeder. But the peace and relaxation can start to turn sour when the food you bought for the birds fattens up the squirrels instead. There is a simple, cost-effective way to keep squirrels off the food.

A MATTER OF TASTE

Birds have a sense of taste, but it is fair to say that their palate isn't as discerning as our own or that of squirrels. They have far fewer taste buds than a squirrel, or most other vertebrates for that matter. Birds, depending on the species, can have fewer than 50 or up to 500 taste buds, whereas we in comparison have 9,000–10,000, so over 20 times as many!

We can use the birds' lack of taste to our advantage. The capsaicinoids in chilli, which cause the burning sensation, are not felt by birds. By including chilli, you will put the squirrels off the food, but leave the birds unaffected.

Spice up your seed

YOU WILL NEED:
- CAYENNE PEPPER
- BIRD SEED
- BIRDFEEDER

Simply mix a small amount of cayenne pepper in with
the bird seed and give it a good shake before putting
it in the feeder. You shouldn't need too much, but you
may have to experiment with quantities.

Keep it clean

Wild birds, like all living things, are susceptible to diseases, and
at feeding stations birds come into close proximity with each other.
You can help them by regularly cleaning your birdfeeders with mild
disinfectant, rinsing with clean water and drying before putting
them back out for the birds.

14 Use reflections to scare off squirrels

** fairly simple   squirrels

Squirrels are tenacious creatures and can be very difficult to outsmart if they are determined to raid your garden's natural larder.

SCARED OF THEIR OWN REFLECTION

Most people would be quite offended if it was suggested that their reflection in the mirror would be enough to scare themselves silly. But this is actually true of many squirrels and a mere glimpse of themselves is enough to make them run away with their tails between their legs.

MIRROR DESIGN

It is not just about scaring squirrels. An added bonus of using a mirror is that it can aesthetically enhance the design qualities of a garden. They create a sense of space and give additional light. Dress the mirror with branches or foliage from a nearby shrub or climber so it doesn't look too stark and out of place.

POSITION WITH CARE

While very rare, in extremely hot conditions, it is possible for an outdoor mirror to cause a fire if it heats up enough and there is dry material such as straw or dead leaves nearby.

Sparkling CDs

Hang your old CDs in the branches of trees where squirrels have been stealing nuts, berries and seeds. The reflections and flashing lights as the sun's rays catch the shiny CDs will spook the squirrels and cause them to flee. You can also hang them from lines of string stretched above your fruit and vegetable crops for the same effect.

DON'T BE DAZZLED

Don't place mirrors on a south-facing wall, as you might glance into directly reflected sunlight, which can damage your eyes. Instead, attach them somewhere cool and shady.

A TACTICAL APPROACH

Watch to see the approach path of where the squirrel is entering the garden. Place the mirror in these areas, keeping them low, to avoid birds flying into them.

Weatherproof your mirror

Although you can buy acrylic mirrors that are suitable for hanging outdoors without any proprietary work, they can be prone to scratching. Interior mirrors require mirror-edge sealer to be applied around their edges, otherwise moisture can seep into these edges, causing them to come away from the backing board. The edges can also blacken due to harsh weather elements such as the cold, damp and frost.

YOU WILL NEED:
- MIRROR
- MIRROR SEALANT (AEROSOL OR LIQUID)
- PAINTBRUSH
- CLOTH
- DENATURED ALCOHOL OR GLASS CLEANER

Place your mirror on the ground and then, depending on whether you have an aerosol or liquid, either spray or paint with a brush around the edges of the mirror. Wait for it to dry and then use a cloth and denatured alcohol or glass cleaner to remove any excess sealer from the surface of the mirror. Once the sealer has dried it can be hung outside.

15 Send squirrels on a slippery slide

✳ ✳ fairly simple   squirrels

Squirrels think nothing of climbing a birdfeeder pole for the promise of a free meal. Beat them at their own game by greasing the pole and watching as they slide back down.

MASTERS OF AGILITY

Anyone who has spent time watching squirrels in their garden will attest to the acrobatics that these mammals can perform – they are worthy of a position in the Cirque du Soleil! Their padded feet can cushion jumps of up to 6m (20ft) and they have been known to run an impressive 20mph (32kph). Not only that, they are formidable climbers, and this includes climbing up birdfeeder poles to take the food you have put out for the birds. But you can be one step ahead of them – by greasing the feeder pole.

SLINKY LOVERS

Another, less messy, option is using a slinky. Slinkys aren't just children's toys – they're also squirrel deterrents. Slide the slinky down the feeder pole, so the squirrel has to navigate it to reach their desired food reward.

Grease the pole

YOU WILL NEED:
- GREASE (PETROLEUM JELLY WILL DO)
- PAINTBRUSH

Apply petroleum jelly to the pole using a paintbrush to avoid getting too messy. It is waterproof and will last for quite a while before you have to reapply.

Grease isn't good for bird feathers, so make sure to only grease the pole and not the feeder or the areas that the birds will come into contact with.

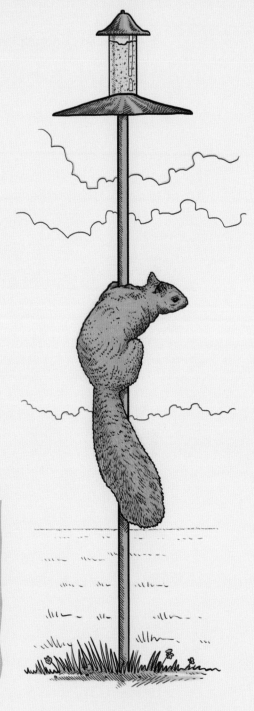

Did you know?

Grey squirrels' (*Sciurus carolinensis*) feet can face forwards and backwards, owing to their double-jointed ankles – no wonder they're such good climbers!

51

16 Keep rodents away from your garden

** ** fairly simple   🐀 rats and mice

Feared by many, loved by few, mice and rats can cause problems for gardeners as they eat their way through seeds and crops, while often nesting in compost bins or the potting shed.

WHAT'S ON THE MENU?

There are probably more rats and mice in the nearby vicinity than you realize. Both are shy, mainly nocturnal creatures, keeping out of sight for most of the day. You may start to notice them if their numbers increase rapidly due to a source of food or warmth, or if they start to damage your garden crops.

Both rats and mice will happily munch through many of your harvests. A favourite of rats is sweetcorn (*Zea mays L*), but they will usually devour most fruit and vegetables, as well as chewing through bulbs and corms.

Mice will also remove seed freshly sown outdoors, and from pots and seed trays in the greenhouse. Their favourite is usually pea (*Pisum sativum*) and broad bean (*Vicia faba*), but they will happily help themselves to whatever seed is available.

MOUSE-PROOF YOUR GREENHOUSE

Mice are tiny creatures and will enter greenhouses and potting sheds through the tiniest of holes. Secure your greenhouse by plugging any small gaps with stones, pebbles or steel wool. Avoid blocking holes with wood or cardboard as mice might eventually chew through them. If possible, have a concrete floor, as rodents can burrow under wooden floors and chew their way through.

FENCE THEM IN

If sowing seeds in pots, place a small wire mesh of around 6.5mm (¼in) over the top of the pot to prevent mice from stealing them.

BIRDFEEDERS

More favourite treats for rats are the seeds and nuts beneath birdfeeders. Only put out enough food for the birds during the day and take it inside at night. Sweep beneath the birdfeeder every night.

SEEDY INGREDIENTS

Think about what you are feeding birds. A lot of seed mixes contain high cereal grain content, which few birds eat (mainly just house sparrows and pigeons). Many others discard these bits to get to the good stuff, which leads to more waste on the ground, waiting for rats to come along. Instead, more expensive things like sunflower hearts (*Helianthus annuus*) lead to less waste and fallen food, and most birds love it. See overleaf for more ways of deterring rodents.

Birdfeeder tip

Position birdfeeders above patios or tarmacked areas. Avoid placing them over grass, as this is almost impossible to clear up each evening, which will then attract rats.

HEAVENLY HEAP

Rodents in compost heaps is a common headache for gardeners. Rats and mice love these popular recycling facilities because they are filled with all their creature comforts. They are warm, as the decomposing plant material gives off heat. They are also dark, and the loose soil material is easy to burrow in. Most gardeners regularly add their kitchen waste to their heap, providing the rodents with regular meals, meaning they don't have to leave their home, as they have everything they need. In short, living in a compost heap is like staying in a rodent five-star hotel!

STARVE THEM OUT

Mice and rats put their stomach before anything else, so if you stop feeding the compost heap with their favourite titbits, they will go elsewhere in search of food. Products to avoid adding to a standard compost heap include eggshells, dairy products, meat and fish.

OPT FOR A HOTBOX

Some compost heaps are designed to get up to super-hot temperatures, such as a hotbox. Although rodents like it warm, the temperatures inside these structures are too hot for even the most stoic rat. They are so hot that it is possible to add meat, fish, dairy products and eggshells, as they will decompose quickly. An additional benefit is that these compost units are rodent-resistant, meaning they make it difficult for anything other than the compostable material to get in.

Clean up

Rats often dwell under decking in the garden, so always clear up your crumbs and food scraps after al fresco dining. If the problem persists, consider a patio instead.

RODENT-PROOF YOUR COMPOST HEAP

If you have one of the plastic, cylinder-style compost bins – often referred to as a 'dalek' – then they can easily be made rat-proof.

The most common way that rats enter the compost heap is by burrowing up through the ground. There are two things you can do to prevent this:

- Place the compost bin on a solid surface such as a patio slab or tarmac. Assuming the remainder of the bin doesn't have any holes in it, rats will find it impossible to get in.

- Place the bin on a fine wire mesh, and cut around it leaving a 5cm (2in) surplus. Use a staple gun to attach the surplus mesh with staples to the outside of the bin. Put the compost heap directly on the soil, so worms in the ground can move in and out, aerating it and speeding up the process of decomposition.

17 Ward off mice with peppermint

✳✳✳ effort required   🐁 mice

If mice are running amok in your potting shed, there is a natural remedy that is easy to implement and will leave the vicinity smelling beautiful.

FOLLOW YOUR NOSE

Mice have fairly poor eyesight, and therefore rely on their strong sense of smell. Growing something pungent tends to deter them from entering a building or invading an area of a garden, as it makes them feel vulnerable due to the fact they can't detect their predators so well. It is known that mice have a particular aversion to the powerful aroma of peppermint (*Mentha* × *piperita*) for this very reason.

TACTICAL PLANTING

One way to guarantee the smell of peppermint in your garden or potting shed is to grow the plant yourself. Peppermint is a herbaceous perennial that is often used as a culinary or medicinal herb. As it is quite invasive – spreading through flower borders if left to its own devices – it is best to grow it in pots to curb its expansive tendencies. This is ideal if you want to place it around the potting shed to keep these tiny rodents at bay.

FIND A FRIEND

A handful of mint plants can be purchased cheaply from garden centres and easily propagated by division to produce lots more. However, if a friend has peppermint plants in their garden, they can supply you with an abundance of free plants.

How to divide a mint plant

YOU WILL NEED:

- TROWEL
- MINT PLANT
- SECATEURS
- 9CM (3¹/₂IN) POTS
- POTTING COMPOST
- SAUCERS

Using a trowel, dig up healthy clumps of mint plant. Use secateurs to lightly trim the roots so the plant will fit in the pot. Place the mint plant in a pot and pour in compost around the sides of the roots, packing it down firmly. Trim back straggly foliage, retaining just a few leaves, so the plant is nice and compact.

Keep the pot on a saucer to capture excess water flowing from the drainage holes and then water the compost. Position pots at roughly every 30cm (12in) around the edges of walls or among potting benches. Don't forget to regularly water these plants.

Essential tip

If you don't fancy growing peppermint, an alternative is to purchase peppermint essential oil. Roll up cotton wool into the size of golf balls and place a couple of drops on each one. Position them strategically around areas where mice might take seeds from.

HOW TO WARD OFF WASPS AND OTHER AIRBORNE IRRITANTS

Ranging in size from large
herons to tiny aphids, aerial garden
visitors may pay you a flying visit
and leave you with damaged crops.
Outmanoeuvre these winged wonders
to ensure you enjoy your harvest
before they do.

18 Build a fruit cage to discourage birds

✳✳✳ effort required 🐦 birds

Soft fruits are delicious, but the problem is, birds think so too! One of the most traditional ways of protecting your soft fruits from ravenous birds is to use a fruit cage.

FRUIT FRENZY

Many birds flock to our gardens in the autumn and winter to take advantage of the bounty of berries on offer. The fleshy, nutritious and tempting fruit encourages these winged seed dispersers to ingest them, and then scatter their seeds far and wide. Fruits are a useful energy source for birds when other food, like insects, is in short supply.

If you really don't want to share the fruits of your labour with our feathered friends, one of the most effective ways to protect them is to make a fruit cage.

CAGE CONSTRUCTION

Making a fruit cage is an investment worth doing well, as the semi-permanent structure can be used for many years. Alternatively, a good option is to purchase one. They are usually made from aluminium and come in all shapes and sizes to suit the extent of your fruit or vegetable beds. The best types are the walk-in ones, as it saves having to bend down to access the crops. These usually have a door on them – just remember to close it when you leave!

Remember to leave your fruit cage unroofed in winter to avoid snow damage and during flowering to allow pollinator access.

Make your own fruit cage

A temporary, cheap structure can be built out of bamboo canes. Here are instructions on how to make a 2m (6¹/₂ft) wide x 2m long x 1.8m (6ft) high cage.

YOU WILL NEED:

- EIGHT 2M BAMBOO CANES (FOUR FOR CROSS BARS, FOUR FOR UPRIGHTS)
- GARDEN TWINE
- SCISSORS OR SECATEURS (TO CUT GARDEN TWINE)
- NETTING
- TENT PEGS
- BLACK POLYTHENE

Mark out the shape of the cage by laying four bamboo canes on the ground in a square. Push the four uprights into the soil, so that 1.8m is above the ground. Use garden twine to attach the four cross bars to the uprights. Pull the netting over the structure until it is taut, and peg the bottom tightly at the base to keep birds and hedgehogs safe. Cover surplus netting with black polythene to prevent weeds that soon grow into netting making it difficult to remove without tearing. Dismantle as soon as the crop is over.

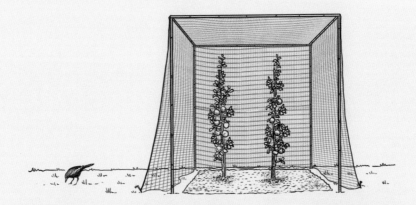

Top tip

Instead of using twine to attach the posts together, you can purchase plastic multi-hole connectors or plastic balls with ports in them, which are used to join bamboo canes together.

19 Protect your fish from herons

** fairly simple herons

While we all love watching birds frolicking in the garden, there is one visitor that might not be particularly welcome if you have a pond. Here are some tricks to save your fish from herons.

HERONS BE GONE!

Herons are large birds that feed on fish. If you have fish in your garden pond, they can quickly disappear if a heron regularly visits. These birds will stand for hours if left undisturbed at the edge of a garden pond, using patience and stealth to catch their prey. Once a fish swims beneath them, their long neck and lightning speed enables them to swoop down and pick them out.

ALSO ON THE MENU...

Although herons mostly eat fish, they will also feed on amphibians and small mammals, plus sometimes reptiles, insects, crustaceans, molluscs, worms and other birds.

Did you know?

There are more than 64 different species of heron, and they appear on every continent around the world, except Antarctica. Some species attract fish by dropping seeds or fruit into the water as a lure.

Give your pond protection

Some of the below suggestions might not be effective in isolation, but combine a few of these to give your fish a better chance of survival.

- **DIG DEEP**
  Herons generally won't go deeper than about 60cm (24in). That does not mean your entire pond needs to be deeper than that, but create deep pockets where some fish might be able to find refuge.

- **SET A NET**
  Placing netting over a pond will prevent herons from being able to stand in the water and swoop in to take fish. Regularly check the net to ensure no other creatures get caught in it.

- **INSTALL A FOUNTAIN**
  Adding a splashing water feature or fountain creates moving water, making it harder for herons to seek out their victims below the surface.

- **GROW WATER LILIES**
  Not only do water lilies look beautiful on the surface of a pond, but their broad floating foliage also provides shelter and hiding places for fish, making it much harder for herons to find them.

- **SCARE WITH A DECOY**
  Herons are solitary creatures and it is believed they will not visit a pond if they think another one is there, so place a fake heron at the water's edge.

20 Frighten off birds with a scarecrow

✳✳✳ effort required   🐦 birds

If you want to stop birds from munching on your seeds or crops, why not follow in the footsteps of many before you and make a scarecrow?

AN ANCIENT CONCEPT

Scarecrows have been used for thousands of years. The earliest records date back as far as ancient Egypt, where wooden structures were erected in wheat fields to protect crops from quail.

KEEP THEM SPOOKED

Birds must consider a scarecrow to be a potential threat for this deterrent to work, and they are likely to get brave over time. If you want your creation to have a long-lasting impact, you'll need to do the following:

Movers and shakers Regularly move the scarecrow into different positions around the garden. Any element of movement can be useful, like clothing that flaps in the breeze.

Change the look Make it look different by simply changing its hat.

Dazzle them Reflective strips or CDs hung from the scarecrow can dazzle and scare the birds.

Other fright tactics

Birds are always vigilant for new threats and often a silhouette of a bird of prey will be enough to keep them away. You can buy kites in the shape of these raptors which, like the traditional scarecrow, is a humane and safe deterrent.

Make your own scarecrow

Scarecrows are easy to make; here's how to
make your own.

YOU WILL NEED:
- TWO LENGTHS OF SPARE WOOD
- WIRE, STRING OR A HAMMER AND NAILS
- BILL HOOK
- OLD CLOTHING, INCLUDING A HAT
- STRAW, OLD RAGS, LEAVES, WOOD CHIP OR
  GRASS CUTTING FOR STUFFING (OPTIONAL)
- HESSIAN BAG OR BUCKET (OPTIONAL)

Fix two lengths of wood (preferably untreated) into
the shape of a cross, which forms the outline of a person.
You can do this with wire, string or a hammer and nails.

Make a point at the end you intend to push into the
ground using a bill hook or similar tool. Add some old
clothes and a hat, and there you have it – your own
bird deterrent!

There are no set measurements, so you can let your
creative juices flow. Use stuffing in the form of
straw, old rags, leaves, wood chip or grass
cuttings to fill out the form and make
it look more realistic. For the
head you could use a
stuffed hessian sack,
or even a spare bucket.

21 Keep wasps away while you dine al fresco

✳ ✳ Fairly simple   🐝 wasps

Don't spend your summer afternoons swiping at pesky wasps and risking getting stung – try this clever method to draw them away from the places where they're being the most irritating.

UNINVITED GUESTS

There is nothing more annoying than wasps buzzing around when you are sitting outside trying to enjoy a meal on a warm summer's day.

To add to the irritation, wasps can become aggressive and deliver a painful sting, particularly later in the season when they are occasionally 'drunk' from feeding on fallen, partially fermented fruit. In addition to ending up in your drink, they can also cause trouble in the garden by damaging ripening fruit.

While some wasp types have a few benefits to gardeners, such as predating on caterpillar eggs, too many wasps can become a nuisance.

SEND THEM ON THEIR WAY

One thing you can do to try to reduce wasp numbers in the garden is to create a 'wasp trap'. These are easy to make, cost practically nothing and will attract wasps away so you can eventually release them elsewhere.

Don't get stung

Take care when releasing trapped live wasps. If too much sugary drink is used and they drown, the corpses release a scent that attracts others.

How to make a wasp trap

YOU WILL NEED:
- 2-LITRE (34FL OZ) PLASTIC BOTTLE
- SCISSORS
- KITCHEN PAPER
- OIL OR PETROLEUM JELLY
- TAPE
- STRING
- MEAT SCRAPS OR SUGARY DRINK

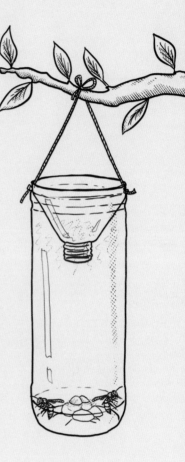

Take a 2-litre (34fl oz) plastic bottle and cut the top 5cm (2in) away, including the nozzle, using a pair of scissors. Remove and discard the lid.

Use kitchen paper to rub oil or petroleum jelly on the inside of the bottle and the removed top section. This will stop the wasp crawling out, once it has entered the bottle.

Invert the removed top section of the bottle, including the nozzle, and place it back inside the bottle. Use tape to hold the inverted nozzle in place at the top.

With the tip of the scissors or a skewer, create two holes on either side of the bottle near the top, then thread string through them.

Fill the bottom of the bottle with bait. Generally, wasps are more interested in meat scraps (protein) in spring and early summer. Later in summer and autumn they prefer sugary drinks for energy. Fill to a depth of 1cm (^{1}/2in) – any more than this and the wasps could drown.

Place the trap away from where you intend to dine. The idea is to tempt the wasps away from you, and towards the trap. Hang the trap from the branch of a tree or fence post.

Check the trap every few days. You may have to occasionally re-grease the inside of the bottle.

22 Hang pheromone traps to minimize maggots

✳ quick fix   ▲ codling moths

Maggots in apples are usually caused by the codling moth (*Cydia pomonella*). By hanging pheromone traps in trees you can reduce the number of egg-laying females, meaning you may never have to bite into a maggoty apple again.

DEFILED FRUIT

We all know that horrible experience of plucking a juicy, plump apple (*Malus domestica*) from a tree, biting into it and discovering it is riddled with maggots. The maggots or caterpillars that do the damage are laid by the codling moth. They can also cause damage in pears (*Pyrus*), quince (*Cydonia oblonga*) and walnuts (*Juglans*).

SPOT THE SIGNS

Before biting into an apple, look for the tell-tale signs that your apple could be maggoty. There is usually an exit hole from the caterpillar near the apple's eye, at the opposite end of the stalk. If you spot a hole, think twice before biting into the fruit.

To make doubly sure, use a knife to cut the apple in half. If codling moths have been present, you will see tunnelling inside the core. There may be frass, which is basically caterpillar excrement. You might even spot the white, brown-headed caterpillar.

CON THE CODLING MOTH

Purchase a moth pheromone trap and hang it in the branches of the trees in early May. These open-sided boxes have a synthetic version of a female pheromone scent (sex chemical), which lures in males in the hope of mating. The bottom has a sticky sheet, which males become stuck on. With fewer males available, the females have a lower success rate of mating, resulting in fewer fertile eggs being laid, and therefore very few maggoty apples.

Although the traps are designed to prevent birds entering and getting stuck on the sticky card, also attach bird netting to the entrances to avoid a bird accidentally getting trapped.

Did you know?

There is a moth similar to the codling moth that affects plums (*Prunus*), called the plum moth (*Grapholita funebrana*). They can be treated in a similar fashion, using a pheromone trap to catch the males.

23 Sow late to cheat carrot fly

** fairly simple   carrot fly

Carrot fly (*Psila rosae*) can completely ruin your crop of carrots if you are unlucky enough to have an infestation in the garden, but sowing seeds later in the season can avoid the problem.

DIVERSE APPETITE

Don't imagine that your other crops are safe from carrot fly – this tiny fly also has a penchant for celeriac (*Apium graveolens* var. *rapaceum*), celery (*Apium graveolens*), parsley (*Petroselinum crispum*) and parsnip (*Pastinaca sativa*).

DAMAGE DONE

It is actually the larvae of carrot fly that inflict the damage. They leave browny-orange scar rings around the carrot, causing them to rot. If you cut a carrot open, it will reveal a network of tunnels caused by these slender, yellowish maggots (the larvae), which are just under 1cm (^{1}/$_2$in) long. This renders the carrots completely inedible.

← Using insect-proof mesh suspended on hoops over carrot plants is by far the best way to control carrot fly.

How to avoid carrot fly

There are a number of tricks to dodge carrot fly infestation.

● **SOWING TIME**

Try sowing carrot seed slightly later in the season. The first generation of carrot fly are usually about prior to mid-May, so if you sow after this date, you should hopefully avoid any damage. Likewise, if you harvest before late harvest, you might avoid the second generation. This traditional gardening technique usually works, but climate change has meant seasons and carrot fly infestations are fluctuating.

● **COVER UP**

Insect-proof mesh suspended on hoops is an effective method of control (pictured opposite). Bury the edges and keep in place until November.

● **CROP ROTATION**

Carrots should be grown in a new spot each year and covered with mesh as soon as the seed is sown. Growing in fresh soil will help keep them free from disease and therefore more resilient to attacks by pests.

● **THINNING OUT**

Carrot fly love the aroma of carrots, so take care when thinning out seedlings, as this releases their scent. Try to sow carrots sparsely to avoid having to remove too many of the surplus carrots. There are handheld seed distributors that help tiny seeds being distributed.

24 Prevent pears from falling early

** fairly simple pear midge

It is disappointing to see young pears on the ground before ripening. Here are some tips on how to avoid pear midge (*Contarinia pyrivora*) from causing your pears to drop early.

PEAR DROPS

Young pears (*Pyrus*) turning brown, looking distorted and dropping prematurely to the ground before they are ripe is usually down to pear midge or frost damage. It is easy to tell which has caused it – simply open up the immature pear to inspect it. If there are maggots in it, then it is pear midge. If there is nothing, then it's frost damage.

KNOW YOUR ENEMY

Pear midge can wipe out a year's crop if there is a bad infestation. Adults emerge from the soil in mid-May, gather around pear trees and mate. The fertile female lays her eggs on the blossom and the maggots form and start to feed on the inside of the young pear. This causes the pear to drop onto the ground where the maggots burrow into the ground and overwinter in the soil before emerging as adults again in April, and the cycle starts again.

BREAK THE CYCLE

If you understand the cycle of pear midge, you can break it – this is key to stopping them reproducing and reduce the infestation. The trick is to lay a weed-suppressing membrane or a tarpaulin below trees in late spring as the young fruit emerges. When the fruit falls to the ground, the maggots won't be able to bury into the soil and emerge the following year. Although you won't save the current year's crop, you can reduce damage to future crops.

GET DIGGING

As an extra precaution, lightly dig over the ground in autumn or winter, taking care not to damage the tree roots. This will expose any remaining cocooned maggots, which are usually around 5–8cm (2–3in) deep. The cold winter will kill off any left near the surface, and some birds will feed on them too. If you have chickens, let them peck over the ground (see page 100).

Pear midge takeaway

Pick up any fallen fruitlets (immature or small fruit) and remove them off-site. Don't add them to the compost heap, as the maggots may overwinter there and then emerge to start their cycle again.

25 Keep your cabbage clear of root fly

✳ quick fix   🪰 cabbage root fly

If you're wondering why your cabbages have stunted growth, look yellow, unhealthy and wilting, they have probably been attacked by cabbage root fly (*Delia radicum*). Here is how to avoid the problem.

BRASSICA BANQUET

Cabbage root fly can decimate any of your plants belonging to the brassica family. This includes broccoli (*Brassica oleracea* var. *italica*), Brussels sprouts (*Brassica oleracea* var. *gemmifera*), cabbages (*Brassica oleracea* var. *capitata*), kale (*Brassica oleracea* var. *sabellica*), oriental greens (*Brassica rapa*), radishes (*Raphanus sativus*), swede (*Brassica napobrassica*) and turnips (*Brassica rapa* subsp. *rapa*). They cause the most harm to seedlings and other young plants, where the larvae (a small, white maggot) tunnel and feed on the roots; older plants can tolerate the harm.

After the larvae has had its fill feeding on brassica roots, it cocoons itself underground and then emerges a few weeks later as a fly. The adult female looks a bit like a housefly (*Musca domestica*). The fly is attracted by the aroma of brassicas and lays its eggs on the soil at the base of stems of young plants. These then hatch as larvae and the cycle continues.

Did you know?

The cabbage root fly has three generations, which appear between late spring and late summer. Young seedlings and small plants are most at risk as they are unable to sustain themselves once their roots have been devoured by the larvae.

CREATE A BARRIER

The best way to solve a cabbage root fly problem is to prevent it laying its eggs at the base of the stem. You can purchase collars from garden centres or online, which fit around the stem, covering the surrounding soil. It is also easy to make your own. Cut 8cm (3in) square or circle collars from carpet underlay or roofing felt. Make a slit from the edge to the centre and slide the collar around the stem. You can also weight it down with pegs or stones.

MORE TRICKS

Here are some other methods to prevent an infestation of cabbage root fly:

Cover up Place an insect-proof mesh over plants from the cabbage family to prevent damage, and bury the sides in the soil. This may be your only option for root brassicas such as radishes, turnips and swedes, where collars are impractical.

Invite predators Encourage natural predators to hoover up any larvae near the surface of the soil, such as ground beetles, birds (see page 97) and hedgehogs (see page 102).

26 Grease bands to stop winter moth caterpillars

** fairly simple   ⌃ winter moth caterpillars

Winter moth caterpillars (*Operophtera brumata*) affect lots of tree types, eating leaves and causing fruit to distort. The key to reducing tree damage is to prevent the wingless female crawling up the trunk.

MUNCH FEST

Winter moth caterpillars affect dogwood (*Cornus*), elm (*Ulmus*), hawthorn (*Crataegus*), hazel (*Corylus*), hornbeam (*Carpinus*), oak (*Quercus*), rose (*Rosa*), sorbus (*Sorbus* subg. *Sorbus*), sycamore (*Acer pseudoplatanus*) and many more trees and shrubs.

There are a few species of winter moth caterpillar, but they all essentially inflict similar damage to ornamental and edible trees. This usually involves creating holes in leaves, blossom and in fruit such as apples (*Malus domestica*), cherries (*Prunus avium*), pears (*Pyrus*) and plums (*Prunus*). This can restrict their growth, and in severe cases may cause a tree to die. The fruit can end up inedible.

Multiple moth types

Winter moth is a generic name for a few species that include the mottled umber moth (*Erannis defoliaria*) and March moth (*Alsophila aescularia*). In all species, the wingless female emerges from the soil and travels up the trunk to lay eggs in the branches between November and April.

WHERE TO TAKE CARE

Ornamental trees can usually survive an attack from winter moth caterpillars. They are very much part of the garden ecosystem, and birds such as tits enjoy feeding on them, so unless there is a major problem, they should be tolerated. On fruit trees, though, crops can be completely wiped out, so you need to take action.

GREASE UP

To help break the life cycle of a winter moth caterpillar, place a grease band or barrier glue around tree trunks in October. These can easily be purchased from garden centres or online. The sticky barrier thwarts the moth's ability to reproduce by stopping the wingless female from emerging from the ground during winter and crawling up the trunk to lay eggs. Make sure the grease band remains sticky until April, and grease any stakes too.

Encourage natural predators such as nesting birds into the garden to feed on the caterpillars (see page 97).

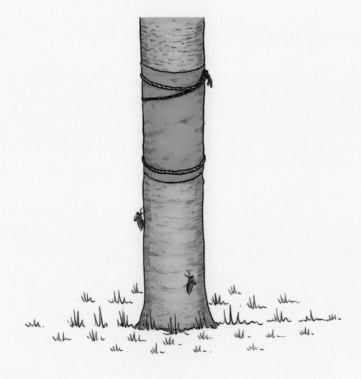

27 Lure bats by inviting in their dinner

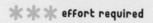 **✳✳✳ effort required**   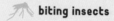 **biting insects**

One way to invite bats into your garden is to plant to attract less annoying invertebrates and moths. Bats are a natural predator and will guzzle both these as well as biting insects – and at the same time will provide you with dazzling night-time aerial displays.

BLIND AS A BAT?

Bats are certainly not blind, as is often thought, but as they come out at night they use echolocation – high-frequency calls that rebound off surfaces and travel back to their ears, building a detailed picture of their environment. This allows bats to catch their insect prey in darkness. Some are fast aerial hunters and others are well adapted to glean invertebrates from foliage. By planting to attract their dinner, you can encourage bats into the garden to hoover up your insect pests and put on quite a show.

LITTLE AND LARGE

Globally, there's a huge diversity of bats. The bumblebee bat (*Craseonycteris thonglongyai*) in Thailand and Myanmar is the smallest mammal on the planet and is, as the name suggests, the size of a bumblebee. Other bats are much bigger and the golden-crowned flying fox (*Acerodon jubatus*) of the Philippines has an impressive wingspan of over 1.5m (5ft).

Did you know?

All bats in the United Kingdom are protected due to huge declines over the last century. By planting to feed bats, you're helping to conserve these winged mammals.

Grow plants that attract insects

- **PALE AND NIGHT-SCENTED PLANTS**
  Night-scented and pale flowers, such as evening primrose (*Oenothera biennis*), night-scented stock (*Matthiola longipetala*) and hemp agrimony (*Equipatorium cannabinum*) can be more obvious to night-flying insects like moths, attracting them into your garden.

Common evening primrose
(*Oenothera biennis*)

- **AROMATIC PLANTS**
  Herbs and aromatic plants can lure insects, so plant things like lavender (*Lavandula*), mint (*Mentha*) and marjoram (*Origanum marjorana*).

- **DIVERSITY OF FORMS**
  If possible, provide plants with short florets, like those in the daisy family, which will provide the insects needed for small bats. Teaming this with plants like honeysuckle (*Lonicera periclymenum*), with long pollen tubes, will encourage a wide diversity of insects, catering for different species of bat.

Marjoram
(*Origanum marjorana*)

Common honeysuckle
(*Lonicera periclymenum*)

28 Beat biting insects by housing bats

✳✳✳ effort required

mosquitoes (and other biting insects)

There is nothing worse than being bitten by mosquitoes and midges when sitting in the garden. Erect a bat box to provide a home for bats that prey on these pesky insects.

KINGS OF INSECT CONTROL

There are more than 1,400 species of bat around the world, some of which help pollinate plants, including species of agave (*Agave americana*), banana (*Musa*) and mango (*Mangifera indica*). As well as helping to provide us with chocolate and tequila, bats can also help with seed dispersal and insect pest control.

All 17 bat species that breed in the United Kingdom only eat invertebrates, from midges and mosquitoes to beetles and moths. Encourage them into your garden and at the same time you will be rewarded with their captivating aerial displays.

← Bats roost in old trees and buildings; as old trees become scarce and buildings made bat-proof, providing roosts is crucial.

Build your own bat box

If you have the time and resources, why not have a go at building your own bat box?

YOU WILL NEED:

- ROUGH SECTION OF NON-TREATED, EXTERIOR WOOD 25MM (1IN) THICK X 150MM (6IN) WIDE X 1,195MM (47IN) LONG
- SAW
- TAPE MEASURE, RULER AND PENCIL
- HAMMER AND NAILS
- RUBBER STRIP
- A TREE TO HANG THE BAT BOX
- ODOURLESS WOOD GLUE

Mark out the dimensions for the box onto the wood as shown in the diagram below, then use a saw to cut the lengths of timber out. The roof and front pieces both need one end cut at a 22° angle (see diagram below).

Roughen up the back of the box by cutting 1mm (^{3}/$_8$in) grooves in it every 10mm (^{1}/$_2$in).

Use the hammer and nails to attach the front, sides, base and back plate together. Allow the back plate to hang 80mm (3in) below the base.

There should be a gap of 15–20mm (^{1}/$_2$–^{3}/$_4$in) left between the base and the back plate for the entrance.

Attach the roof to the back plate by nailing the rubber strip across them. This rubber strip will act as a hinge so the box can be opened (if unoccupied). Any gaps can be sealed with odourless wood glue.

Attach your box to a tree at 5m (16ft) high.

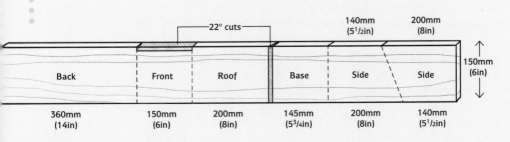

22° cuts

Back	Front	Roof	Base	Side	Side
360mm (14in)	150mm (6in)	200mm (8in)	145mm (5^3/$_4$in)	200mm (8in)	140mm (5^1/$_2$in)

140mm (5^1/$_2$in)   200mm (8in)

150mm (6in)

IT'S BETTER TOGETHER

In the summer, female bats form groups called maternity colonies to give birth and raise their young. A female usually only has a single pup each summer. The males tend to roost in isolation or in small bachelor groups at this time of year. They are nocturnal, so go out to feed at night. By day, both males and females need a safe place to rest, such as bat boxes, cavities and crevices of mature trees, and in buildings.

DO NOT DISTURB

All bats in the United Kingdom are legally protected, so do not disturb an occupied bat box without a licence. Look for dried droppings made up from insect exoskeletons underneath the boxes for evidence of occupancy.

Did you know?

Pipistrelles (*Pipistrellus pipistrellus*) – the bats most often seen in gardens – are so small that they weigh a mere 3–8g ($^{1}/_{10}$–$^{1}/_{3}$oz), which is less than a £1 coin. Despite being tiny, a single bat can eat in excess of 3,000 insects in a night!

Favourite hang-outs for bats

- **CLOSE TO A HEDGE OR TREELINE**
  These linear features help bats to navigate in the dark.

- **IN A WARM SPOT**
  A south-facing aspect is best. If you have more than one bat box, you can provide for different needs by putting up a few boxes facing in slightly different directions.

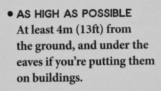

- **AS HIGH AS POSSIBLE**
  At least 4m (13ft) from the ground, and under the eaves if you're putting them on buildings.

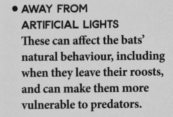

- **AWAY FROM ARTIFICIAL LIGHTS**
  These can affect the bats' natural behaviour, including when they leave their roosts, and can make them more vulnerable to predators.

← The greater horseshoe bat (*Rhinolophus ferrumequinum*) is named for the shape of its nose leaf.

29 Create a ladybird hotel for aphid pest control

✳✳ fairly simple   🐛 aphids

Need an effective, environmentally friendly way to reduce aphid damage? Draw ladybirds into your garden by building them their very own place to stay.

LADYBIRDS TO THE RESCUE

Ladybird adults and larvae love to feed on aphids, particularly greenfly and blackfly. Each ladybird is estimated to consume around 5,000 aphids in their lifetime! An adult ladybird lays 20–50 eggs a day, meaning your colony of ladybirds will continue to breed, providing you with long-term, environmentally friendly pest control.

Having these tiny, spotted beetles in your garden will save you from reverting to spraying to get major pest infestations under control. You can buy ladybird adults or larvae online to release into the garden or greenhouse, or simply encourage them in by creating an inviting habitat for free.

Did you know?

In early paintings, the Virgin Mary, 'Our Lady' was often depicted wearing a red cloak with spots, symbolizing the seven joys and seven sorrows. When farmers in the Middle Ages saw ladybirds eating aphids and saving their crops, they named the tiny beetle 'ladybird' after 'Our Lady'.

IT'S CHECK-IN TIME

One of the best ways of drawing them in is to create a 'ladybird hotel'. Ladybirds seek out dry, safe places in the garden to live, lay eggs and hibernate – places such as hollows of stems or crevasses in dead wood, or even among the scales of pine cones. The nest below replicates their natural environment.

How to make a ladybird hotel

YOU WILL NEED:

- ● FIVE OR SIX PINE CONES WITH OPEN SCALES
- ● STRING BAGS (SUCH AS THOSE THAT PREVIOUSLY HELD ORANGES OR NUTS)
- ● 10CM (4IN) LENGTHS OF DRIED HOLLOW STEMS (SUCH AS BAMBOO, *KERRIA*, *SAMBUCUS* (ELDER) OR *FORSYTHIA*)
- ● STRING
- ● SCISSORS

Place the pine cones into the string bags. Add hollow stems such as short sections of bamboo – poke them through the bags and among the pine cones, then pull the bag closed.

Tie string to the end of the bag and hang it in a sheltered, dry location, such as from the branch of a tree. Pretty soon ladybirds should start to move in, and you will see a reduction of aphid numbers in your garden.

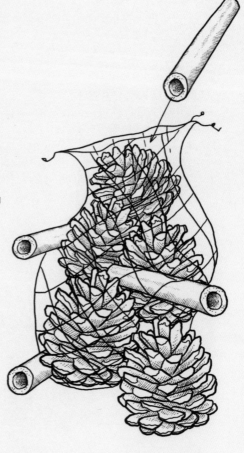

30 Help your veggies thrive with companion planting

*** effort required   aphids and other pests

Pairing two plants in a garden can sometimes benefit one of them by protecting it from pests – this is known as companion planting. Here are some helpful combinations.

A FRIEND IN NEED

It is not just humans and pets that like to be closely surrounded by companions. Believe it or not, some plants enjoy a close relationship with other leafy and flowery friends, where they can share their mutually beneficial qualities to ward off nasty pests.

THE PERFECT PARTNER

Known as companion planting, this popular horticultural method of defeating minibeasts has been practised for centuries. While science is divided on its effectiveness, a few trials have shown there to be some method in the madness. One showed that growing clover in fields of the cabbage family reduced attacks of cabbage white butterflies (*Pieris rapae*) and cabbage root fly (*Delia radicum*), with fewer eggs laid.

Roses and garlic

Plant garlic at the base of your rose shrubs; the pungent aroma from the cloves may help to keep aphids at bay.

Companions to try

Tomato
(*Lycopersicon esculentum*)

French marigold
(*Tagetes patula*)

↑ **DECEIVE THE WHITE FLY**
Plant French marigolds next to tomato plants in the greenhouse, and the white fly could be lured to the French marigolds and stay off the tomatoes.

Carrot
(*Daucus carota*)

Mint
(*Mentha spicata*)

↑ **CONFUSE THE CARROT FLY**
Plant mint close to carrots; the strong scent may confuse the carrot fly (*Psila rosae*), keeping them off the carrots.

Broad bean
(*Vicia faba*)

Summer savory
(*Satureja hortensis*)

↑ DIVERT BLACK APHIDS

Plant summer savory near broad beans; it is believed that summer savory deters black aphids (*Aphis fabae*), which will help the broad beans.

Radish
(*Raphanus sativus*)

Mint
(*Mentha spicata*)

↑ CHEAT FLEA BEETLES

Plant mint close to radishes; it may help to deter flea beetles (*Alticini*), keeping them away from radishes. Plant in pots placed around the radishes rather than directly in the soil, so they don't smother the radishes.

Chrysanthemum
(*Chrysanthemum*)

Chive
(*Allium schoenoprasum*)

↑ USE A DECOY FOR APHIDS

Aphids can be a real problem for gardeners when they start feeding on chrysanthemum shoots and foliage. Try planting a clump of chives next to them – aphids might be deterred by the oniony smell.

Kale
(*Brassica oleracea*)

Nasturtium
(*Tropaeolum*)

↑ TRICK 'CABBAGE WHITE' BUTTERFLIES

Sometimes one plant makes the ultimate sacrifice for their companion. Plant nasturtiums next to your kale, and 'cabbage white' butterflies might lay their eggs on the nasturtiums instead.

CHAPTER FOUR

HOW TO STOP SLUGS AND OTHER MEDDLESOME MOLLUSCS

Safeguard your favourite plants
from the unwanted attention of slugs
and snails. Outsmarting these slippery
customers is easier than you think,
by following a few recommended
tricks of the trade.

31 Drive slugs and snails off your plants

*** effort required

🐌 slugs and snails

Slugs and snails are public enemy number one in the horticultural world. They eat anything from tiny seedlings to large, lush leaves. Here are some tips to save your plants.

SLIMY SABOTAGE

Slugs can wreak havoc in the garden if not monitored. Seedlings can be destroyed within a few hours, while larger herbaceous plants can be defoliated overnight. But there are some easy things you can do to reduce the amount of damage they do.

TIDY UP

Slugs love to hide in nooks, crannies and crevices in the garden. The more pots, trays, slates, rocks, stones or rubbish you have lying around, the more likely slugs will use these places to shelter and multiply, particularly in damp, cool, shady areas. By clearing up any piles of debris, slugs will have fewer areas to take refuge, and birds, toads and hedgehogs are more likely to find them.

Seduce slugs with beer

Did you know that slugs love beer? If you sink a container with a splash of beer into the ground in the evening, slugs will gather there overnight and stay away from your plants. Simply collect the container first thing in the morning and dispose of the slugs elsewhere. Although there is not much evidence to support the use of slug traps or barriers, they will do no harm. RHS research has shown that ferric phosphate pellets and beneficial nematodes (see page 96) are effective.

HAND-PICK

Large slugs and snails are easy to spot and can be picked off by hand. They are very slow moving, so it shouldn't be hard to catch them. Don't forget to wear gloves, as their slime (an external bodily secretion that they produce) can be difficult to wash off your skin. Place them in a container with small holes in the lid and take them to where they will not destroy garden plants. Slugs and snails tend to leave a slimy trail, which should make them easy to track.

CUT BACK

While it is a great idea to have wildflowers and long grass in areas of the garden to attract wildlife, avoid having these features next to vulnerable seedlings, as the damp grass will harbour slugs and snails. Strim or mow areas of long grass near seedlings in the ground.

Plants that slugs love

Choose alternatives to plants that slugs adore, which include:

- Dahlias (*Dahlia*)
- Hostas (*Hosta*)
- Larkspur (*Delphinium*)
- Sweet peas (*Lathyrus odoratus*)
- Tulips (*Tulipa*)

Take extra precautions when growing these in the vegetable garden:

- Beans (*Phaseolus vulgaris*)
- Celery (*Apium graveolens*)
- Cucumbers (*Cucumis sativus*)
- Lettuce (*Lactuca sativa*)
- Peas (*Pisum sativum*)
- Potatoes (*Solanum tuberosum*)
- Pumpkins (*Cucurbita moschata*)

Dahlia
(*Dahlia*)

Larkspur
(*Delphinium*)

Siebold's plantain lily
(*Hosta sieboldiana*)

Cucumber
(*Cucumis sativus*)

Plants that slugs hate

On the other hand, they will steer well clear of these herbaceous plants:

- African lily (*Agapanthus*)
- Bear's breeches (*Acanthus*)
- Black-eyed Susan (*Rudbeckia fulgida*)
- Cranesbill (*Geranium* species)
- Daylily (*Hemerocallis*)
- Elephant's ears (*Bergenia*)
- Foxglove (*Digitalis purpurea*)
- Fuchsia (*Fuchsia*)
- Goldenrods (*Solidago*)
- Granny's bonnets (*Aquilegia* species)
- Houseleeks (*Sempervivum* species)
- Ice plant (*Sedum spectabile*)
- Lungwort (*Pulmonaria* species)
- Masterwort (*Astrantia major*)
- Mullein (*Verbascum*)
- Phlox (*Phlox paniculata*)
- Spurge (*Euphorbia*)
- Wood betony (*Betonica officinalis*)

African lily
(*Agapanthus*)

Black-eyed Susan
(*Rudbeckia fulgida*)

Houseleek
(*Sempervivum*)

Ice plant
(*Sedum spectabile*)

KEEP THEM INSIDE

Small seedlings in the ground are at most risk. They stand a better chance of surviving if they are grown for longer indoors or in the greenhouse before being planted outside. Larger seedlings will have bigger root systems and more foliage, enabling them to endure initial slug attacks. Once the signs of slug damage have been noticed, further action such as hand-picking or nematode treatments may be required to help the plants to continue to survive.

NEMATODES

Nematodes are microscopic worms that can be very effective at controlling slug and snail numbers – you can buy them online. Dilute them in a watering can and pour on soil and plants in areas infested with slugs and snails. You may need to treat infestations regularly. Packets will keep for two weeks after purchase but must be stored in the fridge.

IN BAD TASTE

Like most of us, slugs and snails have their favourite meals – and some they are not so fond of. By growing some of their least-liked plants, it might be possible to thwart an attack.

How does a snail move?

Snails do not simply glide along on their slime as many people think. They walk on a giant foot that extends at the front and pulls the remainder of the foot behind. It's a laborious process, hence the slow speed at which a snail travels, but means they can travel vertically as well as along the ground.

32 Lure birds to restrict pest numbers

✳✳ **fairly simple**　　　　🪱 **aphids, slugs and other mini pests**

Pesky caterpillars, aphids and other invertebrates make a tasty meal for feathered visitors. Encourage birds in by planting to provide them with both shelter and food.

HELPING EACH OTHER

Your garden plants can provide birds with areas to raise their young, to roost in overnight and to supply them with food throughout the year. In exchange for their board and lodgings, they will help reduce the number of pests in your garden, such as aphids, slugs and caterpillars.

TOWN VERSUS COUNTRY

Your location will, in some part, determine the birds that you entice into the garden. House sparrows and starlings (*Sternidae*) are often found in urban gardens, whereas tiny wrens (*Troglodytidae*) and many finches (*Fringillidae*) are most commonly found in rural settings. What is true for all gardens is that you can help attract birds with a few simple planting ideas. See overleaf for the types of plants that will attract birds.

COVER UP

Shrubs and trees offer roosting and nesting opportunities, and also provide cover – plus they are full of insect food and can produce berries or seeds later in the year. Some thorny shrubs like bramble (*Rubus*) and hawthorn (*Crataegus*) can be ideal for birds like blackbirds (*Turdus merula*) and dunnocks (*Prunella modularis*), which nest low in bushes – thorns provide a barrier against predators.

Having an evergreen presence can be advantageous to give cover in the winter, and also in early spring for early breeders. A climber such as ivy (*Hedera*) is a good example of this and may create nesting opportunities for birds like robins (*Erithacus rubecula*) in the spring, but grant cover throughout the year.

SEEDS AND WEEDS

When encouraging birds, be relaxed with your gardening style and allow plants to set seed. Leave cutting back herbaceous borders until early the following year. Sometimes 'weeds' can be good thing. Dandelions (*Taraxacum*) and thistle (*Silybum*) seeds are a fantastic resource for birds. If you do not want to let these run wild in the garden, consider planting field scabious (*Knautia arvensis*), lavender (*Lavandula*), lemon balm (*Melissa officinalis*) or teasel (*Dipsacus*). If you leave some of the seed, you may be rewarded with a nice winter flock of finches.

If you have the space, trees can also offer a great crop of seeds. Alder (*Alnus*) and silver birch (*Betula pendula*) – preferring wet soil conditions – are good options, as they mature reasonably quickly.

PLANT THE BERRY BEST

Anyone who has grown soft fruits can attest to birds' love of berries. One sure way to draw in birds over the autumn and winter is to provide a variety of berry-producing plants. Try blackberry (*Rubus*), elder (*Sambucus*), guelder rose (*Viburnum opulus*), holly (*Ilex*) and rowan (*Sorbus* subg. *Sorbus*) (mountain ash) – and there are many more! Ivy (*Hedera*) provides a fantastic berry resource late in the season, when other berries have gone.

Plants that will bewitch birds

Common ivy
(*Hedera helix*)

Field scabious
(*Knautia arvensis*)

Silver birch
(*Betula pendula*)

Guelder rose
(*Viburnum opulus*)

33 Keep chickens to reduce slug numbers

✳✳✳ effort required    slugs

There are many benefits to keeping chickens in the garden – one of them is that they love to feed on slugs.

FOWL FRIENDS

Keeping chickens is a rewarding hobby. Not only do they provide you with a continual supply of eggs, but they don't take up too much space and make wonderful pets. An additional benefit is that one of their favourite meals is slugs – handy if you have a problem with these menacing molluscs eating your vegetables and herbaceous plants.

There are lots of different breeds of chicken, and some will lay eggs more frequently than others. So if you want an egg for breakfast most mornings, choose your breed wisely.

Did you know?

Hens will still lay eggs without a cockerel. However, the eggs will not be fertilized, so there will not be any chicken offspring without one.

POULTRY CARE

Chickens are generally social creatures, so it is best to have more than one hen – but avoid more than one cockerel (male), as they tend to fight. Chickens need fresh water and feeding once a day – they will enjoy most kitchen vegetable scraps.

Supply them with a watertight and draught-free chicken coop that has a perch, and cover the floor in shavings or straw. Regularly change this bedding and add it to the compost heap. They will need a fox-proof run for them to exercise in during the day.

PECKING ASSISTANCE

Let chickens out onto bare soil for a few days prior to sowing to hunt out slugs; don't let them out when plants or seeds are present, as they are likely to eat these. Chickens will also dig up and feed on any weeds, and will spread their manure, which is an effective fertilizer. Put chickens on the lawn and long grass to reduce the numbers of slugs that might move onto your plants. They will also devour vine weevil that could be damaging soft fruit (see page 136).

34 Recruit hedgehogs to keep slugs in line

** ✳✳ fairly simple ✳✳ **   🐌 slugs and snails

Have a problem with slugs? Then lure hedgehogs into your garden – they love to feed on slugs and other minibeasts, so will help to keep numbers down.

A PRICKLY SITUATION

A hedgehog can be a gardener's best friend. Not only are they cute to look at, but they can also assist with the control of slugs, snails and other small garden pests, as they love to munch on them. Building a simple hedgehog home will not only provide you with a maintenance-free wildlife 'pet' that is a joy to watch, but it will save you hours of time trying to hunt down and remove slugs.

WAIT AND WATCH

Less time controlling slugs allows for the much more rewarding hobby of hedgehog watching. Mind you, they're shy creatures, and will usually only come out between dusk and dawn. They also tend to hibernate for most of winter. But if you are patient, it is very rewarding to see a hedgehog bumbling slowly across your lawn.

Did you know?

Hedgehog numbers are on the decline, so providing them with a home will help them survive.

Make a hedgehog home

YOU WILL NEED:
- WOODEN BOX OR FLOWER POT
- DRY LEAVES, STRAW OR GRASS
- TWIGS

To make a hedgehog home, simply turn a wooden box or large plastic or terracotta plant container on its side. The entrance needs to be at least 35cm (13¾in) in diameter. Fill the lower half with dry leaves, straw, grass or other herbaceous material to act as bedding.

Tuck the home somewhere quiet in the garden, such as at the foot of a shrub or hedge. Avoid areas where pet dogs can pester the prickly mammals. Place soil or twigs over the top and sides to disguise the home and provide additional insulation.

EVERYBODY NEEDS GOOD NEIGHBOURS

In a few weeks' time, you may have a new prickly bundle of joy as your neighbour and an environmentally friendly pest-control buddy. If things go to plan, an infestation of slugs and snails will be a thing of the past.

35 Invite natural predators in with a log pile

** fairly simple   🐌 slugs, snails and hedgehogs

Dead wood is an easy way of providing a rich habitat for many a minibeast and can be fabulous for encouraging hedgehogs and other slug and snail hunters into your garden.

PILES OF DEAD WOOD

In the countryside dead wood is abundant, providing a home to many minibeasts. By building log piles in your garden, you can use your waste materials to encourage a plethora of natural predators, including voracious slugs and snails. For hedgehogs, log piles provide a space to raise young and to safely hibernate in, as well as supplying an all-you-can-eat buffet of tasty insects.

Log piles also provide shelter for slow worms (*Anguis fragilis*) – legless lizards that are often found under wood piles, large stones and in compost heaps. Their backward-curving teeth are just right for gripping on to slippery meals like slugs and snails.

Did you know?

Slow worms have been known to live for up to 30 years in the wild, so are a long-term friend!

MAKE A LOG PILE

Use your woody cuttings from the garden and simply stack them in an undisturbed corner, in shade or dappled shade if possible. There is little maintenance, but as the wood rots down, add a little more.

If you don't have logs, don't worry, as leaf and stick piles can provide first-rate accommodation for a snuffling hedgehog or slithering slow worm. You can also contact a local tree surgeon to provide you with logs.

KEEP THEM COMING

You can also help hedgehogs by putting out water for them in a small saucer, especially in dry conditions. They need to be able to get around from garden to garden freely – they move 2km (1¼ miles) on average in a night – so make sure there is a small hole at the bottom of your fence that they can get through.

36 Attract frogs to feast on slugs

*** effort required    slugs

Ponds are like magnets to wildlife, attracting a rich range of biodiversity. Best of all, they draw in frogs, which feast on slugs and snails.

WELCOMING WILDLIFE

A pond makes an eye-catching feature in the garden. Calm, still water adds a relaxing, soothing atmosphere to an outdoor space, whereas running water or fountains evoke an element of vibrancy and excitement. A pond will also attract frogs – and they will help to reduce slug and snail numbers.

Loads of other creatures will be lured to a wildlife-friendly pond, including hedgehogs, toads, bats and birds, all of which will add to a richer biodiversity and reduce infestations of pests such as midges, mosquitoes and aphids. See overleaf to find out how to make a pond.

Did you know?

The natterjack toad (*Bufo calamita*) (pictured) is extremely rare and is protected by law. Usually found in coastal regions and lowland heaths, it is recognizable by the distinctive yellow stripe down its back. It is also a surprisingly fast runner.

Frog features and toad traits

Common European frog
(*Rana temporaria*)

FROGS
- Frogs have smooth skin.
- They breathe through their skin as well as lungs.
- They have a brown patch behind the eye and come in various shades of green, yellow or brown.
- They move by hopping or leaping.
- Their diet includes slugs, snails, beetles and woodlice.

Common European toad
(*Bufo bufo*)

TOADS
- Toads have dryish warty skin.
- They can live in slightly drier places than frogs.
- They are usually greyish-brown.
- They crawl instead of hop.
- They eat beetles, woodlice and ants.

PLAN YOUR POND

Ponds tend to be difficult to relocate, so give its positioning a lot of thought before starting to dig. A pond's ideal location is on level ground (unless you want to create a waterfall), which is partly in full sun but also has some dappled shade. The shade will help prevent the pond from going green with algae too fast, while the sun will encourage dragonflies, mayflies and many more creatures that will bathe in the sunlight.

How to make a pond

YOU WILL NEED:

- SAND OR HOSEPIPE
- SPADE/PICKAXE (OR MINI DIGGER FOR REALLY BIG PONDS)
- PLANK
- SPIRIT LEVEL
- BUILDING SAND (OR OLD CARPET/NEWSPAPER)
- BUTYL RUBBER LINER
- STANLEY KNIFE TO CUT LINER
- WATER
- ROCKS

Mark out the shape of the pond with either sand or a hosepipe. Most wildlife ponds are curvy and irregular, as this looks more natural. Start to dig the pond and remember to vary the depth in different places, to replicate a natural pond.

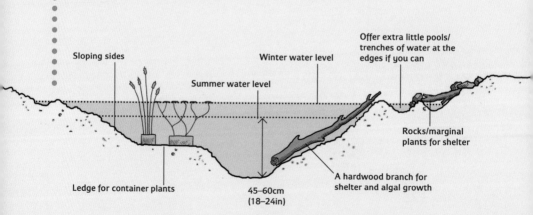

Sloping sides

Winter water level

Offer extra little pools/trenches of water at the edges if you can

Summer water level

Rocks/marginal plants for shelter

Ledge for container plants

45–60cm (18–24in)

A hardwood branch for shelter and algal growth

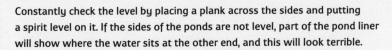

Constantly check the level by placing a plank across the sides and putting a spirit level on it. If the sides of the ponds are not level, part of the pond liner will show where the water sits at the other end, and this will look terrible.

Make sure one side of the pond is shallow to allow small creatures to approach the water without the risk of falling in. Construct a shelf around some of the edges about 10cm (4in) below the surface for marginal plants. Remove any stones or sharp objects that may puncture the liner.

Line the pond with a 5cm (2in) depth of building sand. (You can use old carpet or old newspaper, but sand is best as it is sterile and will not have harboured any undesirable microbes.) Lay a butyl liner over the sand and cut it at the edges, leaving a 10cm (4in) overhang. Dig a shallow ditch on the outside of the pond to bury the edge of the liner.

Fill the pond with water. Rainwater is best, as the nutrients in tap water can turn the water green with algae quite quickly. Place rocks around the edge, creating some overhangs with some of the larger ones. Ensure none of the rubber butyl liner is uncovered, as it can perish quickly if exposed directly to sunlight.

Plants can be added to the pond around 10 days after adding water. To attract a variety of wildlife, choose plants suitable to different depths from fully submerged to marginal and bog plants. Most importantly for frogs, make sure the pond has some water lily pads (*Nymphaeaceae*) for them to sit on and soak up the sun.

37 Toughen up plants to withstand pests

✳✳ **fairly simple**   🐌 **slugs, snails and other pests**

One of the key ways to outsmart garden pests such as slugs, snails, blackfly and greenfly is to build up your plants' resilience. Feeding plants is a great way to do this.

Make your own feed

Nettle (*Urtica dioica*) or comfrey (*Symphytum officinale*) liquid feed is so easy to make.

YOU WILL NEED:
- BIN BAG
- FOLIAGE AND STEMS OF NETTLE OR COMFREY
- BUCKET
- SECATEURS
- WATER
- BRICK
- RECYCLED PLASTIC BOTTLE
- WATERING CAN

Collect a bin bag full of foliage and stems of nettle or comfrey. Cut them up into a bucket in lengths of about 10cm (4in) using secateurs. Fill the bucket with water and place a brick on the plant material to keep it submerged. Leave the comfrey or nettles to decompose fully into the water for a few months. This liquid smells, so keep it away from where you spend time in the garden.

When ready, decant the mixture into a recycled plastic bottle. To use as a feed, dilute it with water at a rate of one part feed to ten parts water in a watering can. You're now ready to feed the root area by watering at the base of fruit and vegetable plants once a week.

IMMUNE BOOST

Healthy plants are far more likely to outlive an attack by insects than unhealthy ones, in exactly the same way that a healthy human can survive illnesses better than weaker people who might succumb to them. By feeding plants a healthy, nutritious and balanced diet, they will have a much higher chance of survival.

LIQUID FEED

Nettle or comfrey liquid feed will provide your plants with much-needed nutrients. Feed made from nettles contains nitrogen and should be fed to plants during the earlier part of the growing season, as it will help them grow and produce lush, green foliage. Once the plant looks as though it is about to produce flowers, it will benefit from a comfrey feed, which is high in potassium. This will help plants produce colourful, large flowers and bumper crops.

CREATE A COMFREY PATCH

Comfrey (pictured below) is a clump-forming perennial that grows to about 80cm (2^{1}/2ft) high. It is easy to grow in a corner of your garden, giving you a supply of free plant food on hand. Plant at 15cm (6in) apart in a grid – about nine plants is enough for a small garden. Keep young plants well-watered in the first year and start harvesting their foliage in the second year to make comfrey feed.

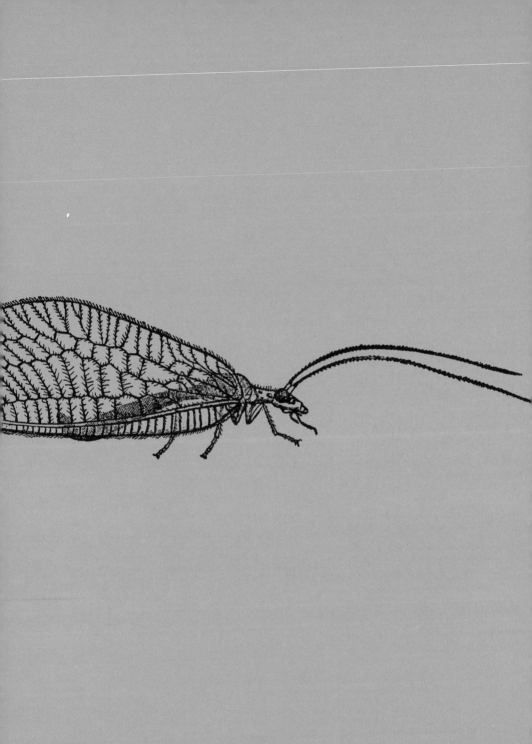

CHAPTER FIVE

HOW TO OUTWIT ANTS AND OTHER INVASIVE INVERTEBRATES

Sometimes it is the smallest of creatures that causes the greatest damage to plants. Learn how to outwit ants, caterpillars, mites, weevils and other tiny creepy-crawlies.

38 Scrub out the woolly aphid

✳ **quick fix**   🐛 **woolly aphid**

Don't be fooled by their fluffy disguise. Underneath their protective coating of waxy cotton wool, sap-sucking aphids are hiding! But you can outsmart them with a simple toothbrush.

WHAT'S THE PROBLEM?

During the summer months you might spot white fluffy clusters in the nooks and crannies of your apple trees (*Malus domestica*). This is not likely to be a fungal infection, but an infestation of woolly aphids (*Eriosomatinae*) (pictured below). Underneath their protective coating they are sucking away at sap just like other aphids, and although they may not be doing serious harm, their activities can cause problems later on, allowing diseases such as canker to develop.

RECYCLE YOUR TOOTHBRUSH

To stop an infestation in its tracks, check apple trees in April for the first signs of infestation and simply scrub off any fluffy clusters that you find. Any stiff-bristled brush will do, but an old toothbrush is perfect for getting into small spaces.

DRAW IN THEIR ENEMIES

Aphids make a tasty meal for many garden visitors. Growing insect-attracting flowers nearby will not only help bring hard-working pollinators into the garden, but it will also help to control aphid numbers generally. Flowering plants like angelica (*Angelica archangelica*), cow parsley (*Anthriscus sylvestris*), oxeye daisy (*Leucanthemum vulgare*) and sea holly (*Eryngium*) are favourites of hoverflies and lacewings, both great aphid controllers, as are ladybirds (see page 84).

Wasps also have a role to play. Smaller parasitic wasps lay their eggs in aphids and bigger wasps simply eat them.

Earwigs may be a problem in other parts of the garden, but can be encouraged to take up residence near your apple trees to control woolly aphids with the same method of an upturned flower pot on a stick (see page 116).

Did you know?

Woolly aphids are most commonly seen on ornamental apple trees, but they also attack *Pyracantha* and some species of *Cotoneaster*.

39 Defend your blooms from earwig damage

** fairly simple   earwigs

To catch earwigs, you need to approach with stealth, luring them into a cosy trap. You'll then be able to humanely remove them and rescue your vegetation from damage.

A TRAIL OF DEVASTATION

Despite being a useful insect that clears up debris in the garden, earwigs can cause untold damage to your prize dahlias (*Dahlia*) or chrysanthemums (*Chrysanthemum*) if left unchecked.

Their favourite trick is to nibble off the young leaves and petals, creating large holes in the plants. Very often – over just a few days if there are enough earwigs – they will strip back foliage entirely to leave just a skeleton of leafy veins. This in time will kill the plant as it is not able to photosynthesize without the green chlorophyll in the leaves to produce its energy for growing.

Five facts about earwigs

- They are nocturnal.

- There are around 2,000 known species worldwide.

- Earwigs have wings, but are reluctant to fly.

- Although they are mostly vegetarian, they will eat carrion and other insects.

- A female will lay around 30–50 eggs and will protect them through winter; she then feeds the nymphs until they are old enough to look after themselves.

Make an earwig trap

You will need to entice earwigs into an attractive trap so that you can move them out of your garden without harming them.

YOU WILL NEED:
- BAMBOO CANE
- STRING
- FLOWER POT
- DRY GRASS OR SHREDDED PAPER

Push a bamboo cane into the ground next to the base of your affected plants. Carefully tie the bamboo cane to the stem using string. This will have the added bonus of providing additional support for the plant.

Fill a small plastic or terracotta pot with dry grass or shredded paper. Then turn the pot upside down and hang it off the top of your cane.

Earwigs love dark, sheltered places, so will soon move into your cosy flowerpot trap. The more traps you make, the more earwigs you will be able to catch.

Check your traps every couple of days to see how many earwigs have moved in, and then relocate them to fruit trees and shrubs, where they will feed on aphids.

Did you know?

Despite their name, earwigs are very unlikely to venture into a human ear!

40 Hand-pick lily beetles to save your flowers

✳ quick fix   🪲 lily beetles

Lily beetles (*Lilioceris lilii*) literally live for lilies: leaves, flowers, seed pods – as long as it is lily-flavoured, these pesky but bright-red beetles will happily munch on it. Here is how to outsmart them.

WREAKING HAVOC

Lily beetles specialize in destroying these most glorious of midsummer flowers and their relatives, turning glamorous blooms into shot-holed, chewed-up disaster zones.

The problem is at its worst when the larvae are born. Although they too are orange-red, these voracious grubs cover themselves with their own black sticky excrement to avoid the attention of birds. They cause damage to almost every part of the plant above ground and can strip it within a matter of days.

The adult beetle with its bright-red colouring is easy to spot, but not so easy to catch, and should never be confused with the useful ladybird.

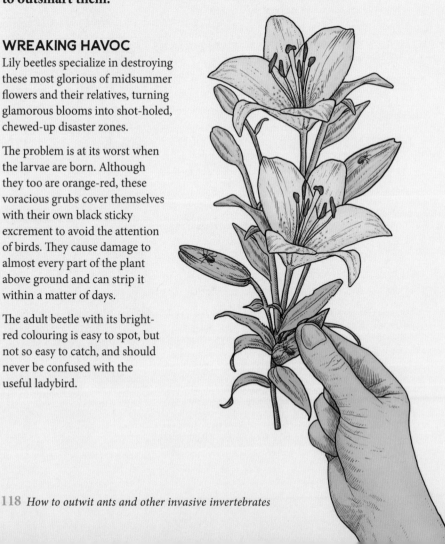

KEEP THEM IN CHECK

Regularly look for adult beetles on plants and pick off any that you see by hand. Dispose of them with food waste or in your green waste bin.

If there are any larvae lurking, scrape or brush off larval blobs by hand and resign yourself to a thorough handwashing afterwards.

HANDS OFF

Lily beetles are experts at dropping through foliage if disturbed, so put newspaper around the bottom of a plant and either shake the stem or tap close to the beetles. Fold the beetles up in newspaper and dispose as you see fit. Alternatively, try a handheld vacuum cleaner.

Adult lily beetle facts

- They are active from late spring through to autumn.

- They are found on members of the lily family including lily of the valley (*Convallaria majalis*) and crown imperial (*Fritillaria imperialis*).

- They are usually visible in the evening or early morning in small numbers.

- Larvae are active in midsummer and the black sticky blobs hiding the larvae are easily spotted.

41 Build a bird table to control caterpillars

*** effect required   ⌃ caterpillars

Caterpillars make short work of munching through leaves, fruits, stems and roots. Encourage birds into the garden to reduce caterpillar numbers by building a bird table.

MINIBEAST BANQUET

If you've had enough of caterpillars' voracious appetites and the damage they cause, one of the best things you can do is invite birds into your garden. Many birds love to feed on caterpillars, and once they have been enticed in with a bird table, they will stay and helpfully reduce caterpillar infestations – they are great fun to watch too.

← The blue tit (*Cyanistes caeruleus*) is one species of bird that might fly into your garden to munch on caterpillars.

Build a bird table

Here is how to construct a feeding table for birds.

YOU WILL NEED:
- UNTREATED WOOD
- BRASS OR GALVANIZED SCREWS
- DRILL
- SAW
- ANGLE BRACKETS

You will need a flat rectangle for your tabletop of 50–100mm (2–4in) thickness; around 30 x 50cm (12 x 20in) is a good size. Add a rim around the edge (about 1cm/$\frac{1}{2}$in in height), leaving a small gap. The rim stops food from being spilt; the gap allows you to easily clear food and adds drainage.

You can put a roof on the table – this looks nice and shelters the food and birds from the elements. Make a cross to form the base, and attach an upright piece of timber for the stand at your desired height. Secure your tabletop to the stand and use angle brackets for additional stability.

Birds need the security of cover, so place the table near a shrub or tree that allows them to retreat if they sense danger. If there are cats in your neighbourhood, avoid low cover where they wait to ambush.

Regularly clean the bird table with a mild disinfectant, rinse it and let it dry before using it again.

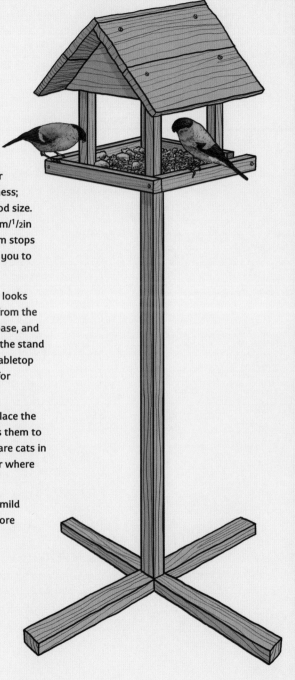

121

ENEMY NUMBER ONE

A caterpillar will find food to sustain itself through the pupal stage and go through a metamorphosis, turning into a butterfly or moth. They certainly eat a lot. Some caterpillars are reported to eat 27,000 times their own body weight!

Many songbirds eat insects in the spring and summer, including caterpillars, which provide a vital source of protein for their young chicks. Research on great tits (*Parus major*) has shown that they are able to synchronize their breeding to coincide with when caterpillars are plentiful. The success of the birds' breeding attempts can come down to whether or not they find enough food.

Songbirds like house sparrows (*Passer domesticus*) and tits will consume vast amounts of caterpillars – one blue tit chick can eat up to 100 each day. Your feathered friends will also eat beetles, dragonflies, aphids and spiders.

FEEDING TIME

A range of foods will attract a wider variety of birds: sunflower seeds (*Helianthus annuus*), high-energy seed mixes and chopped-up peanuts (*Arachis hypgaea*); fruits like grapes (*Vitis vinifera*) and chopped apples (*Malus domestica*) can attract thrushes; small amounts of grated cheese will attract robins and dunnocks. In spring and summer, mealworms are a good (but expensive) option – soak dried mealworms before you put them out. Avoid too much crumbled bread, as it doesn't provide the nutrition birds need.

Bath time

Birds always need water, so a bird bath, refreshed each morning, or a pond are excellent additions to encourage feathered friends into your garden throughout the year.

42 Protect box plants from box tree caterpillars

✳✳✳ effort required   ∿ box tree caterpillar

There has been a huge increase in the decimation of box plants recently due to infestations of box tree caterpillars (*Cydalima perspectalis*). To outsmart this voracious caterpillar, you may have to remove them by hand or select alternative plants.

ATTACK ON THE BOX TREES

Box trees (*Buxus*) are a popular choice due to their small evergreen leaves and compact habit, making them ideal for clipping tightly into shapes for topiary, knot gardens and edging the front of borders. However, the box tree caterpillar will munch through the foliage of many of these plants.

DOUBLE TROUBLE

Box plants have been hit a double blow: it is not just the problem of box tree caterpillars destroying many historic and domestic gardens, they also have a fungus called box blight to deal with. If you have an ailing box plant in your garden, it is important to be able to distinguish between the two potential causes of its demise to get to the root of the problem.

Box blight

Box tree caterpillar damage

BLIGHT OR CATERPILLAR?

Box blight presents as brown foliage, which eventually drops off to leave bare patches and sections of dieback on the tree (see above left). The young stems can also have black streaks on them, and in wet conditions it is sometimes possible to see the white fungus on the underside of the leaf.

Attack from the box tree caterpillar also results in some dieback, but there is usually white webbing in among the foliage and stems around the area where a caterpillar has been feeding (see above right). It may also have stripped the bark and girdled the trunk. If you look carefully, you may spot the distinctive-looking caterpillars themselves, which are greenish-yellow, have a large black head and white stripes down their sides, or the adult moth, which has creamy-white wings with a brown edge.

See opposite for details of how to deal with box blight and box tree caterpillars.

Plant alternatives

If your problems with box tree caterpillars and box blight are insurmountable, the only solution is to look at alternative plants. There are many other small-leaved evergreen shrubs that are not susceptible to the box tree caterpillar. They are suitable for hedging and topiary, and some of the more compact ones can be used for edging knot gardens and potagers.

- Chilean myrtle (*Luma apiculata*)
- Darwin's barberry 'Compacta' (*Berberis darwinii* 'Compacta')
- Delavay osmanthus (*Osmanthus delavayi* AGM)
- English yew (*Taxus baccata* 'Repandens' AGM)
- Fortune's spindle (*Euonymus fortunei* – various cultivars)
- Hebe (*Hebe pinguifolia* 'Sutherlandii')
- Honeysuckle 'Baggesen's Gold' (*Lonicera nitida* 'Baggesen's Gold' AGM)
- Japanese holly (*Ilex crenata*)
- *Pittosporum* 'Arundel Green' AGM
- Rosemary (*Salvia rosmarinus*)

EXTRACTION TECHNIQUES

Handy work The most effective method of getting rid of box tree caterpillars is to pick them off by hand. Due to their bright-green and yellow colour, they are easy to spot. Do this once a day as soon as you see symptoms arise.

Natural born killers You could also use biological control in cool, wet weather – spray the area with a box tree caterpillar nematode, which are natural predators and will help to reduce numbers. Nematode packs can be bought online and should be stored in the fridge for no longer than two weeks until being used.

Lure them in Use a box tree caterpillar moth trap that uses a pheromone lure to attract male moths, where they fall into a funnel trap. Reducing the number of available males from mating with female moths may help to minimize egg laying and reproduction.

43 Rescue blackcurrants from midge attack

✳✳ fairly simple   ⌒ blackcurrant gall midge larvae

This tiny pest can be a real nuisance in the fruit garden, especially to blackcurrant crops. As with many plants, selecting resistant varieties can ensure your crops aren't affected.

MIDGE ATTACK

The blackcurrant gall midge (*Dasineura tetensi*) is a tiny 2mm (^{1}/$_8$in) fly that lays its eggs in the creases of newly emerging blackcurrant (*Ribes nigrum*) leaves in spring. However, it is the hatching larvae (or maggots) that do the damage, feeding on the tops of leaves and shoots, stunting their growth and reducing the yield of crops. They are especially harmful to young plants.

Did you know?

The popular blackcurrant drink Ribena is derived from its botanical name *Ribes nigrum*. Blackcurrant cordial is packed full of vitamin C and is therefore extremely healthy. In Britain during World War II, as there was a shortage of imported fruit such as oranges and lemons, children were given Ribena for free to ensure they were receiving their daily dose of vitamins.

CHOOSE RESISTANT VARIETIES

Varieties of blackcurrants such as 'Ben Connan' and 'Ben Sarek' have good resistance to gall midge and will be unaffected by their presence. Choosing resistant varieties of most crops can help you beat the battle against pests.

Some other examples include carrot varieties that are resistant to carrot fly, including 'Flyaway' and 'Resistafly', and the potato variety 'Charlotte' that slugs are said to steer clear from.

BREAK THE CYCLE

Hoeing the area around blackcurrant bushes during dry periods can help reduce infestations. The theory is that it breaks the cycle as the larvae drop off the leaves after feeding to pupate in the soil, before emerging later as flies. By disturbing the soil, it exposes the pupae, which will either get dried up under the sun or eaten by birds. Just be careful when hoeing not to damage emerging blackcurrant shoots from the stool bush.

44 Stop gooseberry sawfly in its tracks

✳ **quick fix** ⎯⎯ **gooseberry sawfly larvae**

One day your gooseberry bushes look fine. A few days later almost all the leaves have disappeared. This is a sure sign that gooseberry sawfly (*Nematus ribesii*) eggs have hatched into voracious larvae.

SNEAKY CREATURES

Sawfly larvae look a lot like caterpillars (pictured below): either green or green with black spots, depending on the species (see box, opposite). They feast on leaves, methodically eating from the outside in, and often moulding their bodies to the edge of the leaf as a sly disguise tactic, although some rise up to look more threatening when disturbed. They are relentless and can completely defoliate an entire plant at speed.

There can be more than one generation in a single season, depending on which species of sawfly you have, so don't presume that the bush won't be attacked again as it recovers. They may also perform the same leaf-disappearing trick on your currant bushes. While sawfly damage won't kill your fruit bush, it may reduce the yield and gradually diminish its vigour.

BE QUICK

Sawfly lay their eggs on the underside of lowdown leaves from April onwards, so begin regular visual checks early for the first signs of larvae appearing. Pick off any larvae you spot and keep checking and picking throughout the warmer months. Don't forget your gloves!

SOME GOOD NEWS

If defoliation happens late enough in the season when your fruit are fully formed, then at least your fruit (and thorns) are more visible, making the prickly job of harvesting gooseberries a little easier.

Types of gooseberry sawfly

There are three different types of gooseberry sawfly: 'common' (*Nematus ribesii*), 'pale-spotted' (*Euura leucotrocha*) and 'small' (*Pristiphora appendiculata*). All inflict similar damage, but have slightly different life cycles; they also attack red currants and white currants.

45 Stay one step ahead of red spider mite

✳ quick fix   🕷 red spider mite

If you spot mottled leaves and finely spun webbing on a plant, get out your magnifying glass, put on your deerstalker hat and check for the presence of these tiny, red-orange mites.

WEB OF DECEIT

Red spider mites (*Tetranychus urticae*) can infest all manner of plants, particularly in hot, dry conditions. Tiny but prolific, these spider relatives spin fine webs between leaves, making a bridge to travel from place to place to feed. The sheer number of mites feeding cause the mottling effect, eventually affecting the overall vigour of a plant.

HIDE AND SEEK

You will find armies of red spider mites in the following places:

- Glasshouses (especially heated ones) – they can be a problem here all year round.

- On conifers – they usually have to be tolerated here due to tree size.

- On fruit trees.

DRIVE THEM AWAY

Your best strategy is to make red spider mites less comfortable in your garden. Try spraying water over affected plants with a hosepipe after the sun has gone down. Not only will you knock off any number of mites, but the humidity created will discourage them from breeding more heavily. In your greenhouse, regular damping down (spraying or pouring water on paths and surfaces to increase the humidity to at least 65%) will have a similar effect. For minor infestations, spray with a plant-based oil or horticultural soap. Biological control is also available.

If you are lucky enough to have a greenhouse, remember to have a tidy-up in the winter months. Eggs and mites overwinter wherever they find shelter, so remove any debris and old equipment, weed in and around the greenhouse, and disinfect with a horticultural product. Don't give these mites a winter home! On fruit trees you can prune out small infestations.

46 Prune to outmanoeuvre fuchsia gall mite

✳✳ Fairly simple   🕷 Fuchsia gall mite

Infestations of this tiny mite have increased dramatically over the past few years, with many beautiful fuchsia specimens succumbing to them. Outwitting the mite requires stealth and caution.

SPOT THE LITTLE MITE

For a gardener, fuchsia gall mites (*Aculops fuchsiae*) can be very galling indeed. This newly introduced species (probably from South America) is so tiny that it is impossible to spot them in the early stages of an infestation, unless you are the sort of person who examines their plants with a microscope. What is clear to see is the damage they cause as an infestation grows on your prized specimens.

These tiny sap suckers feed on fuchsia foliage, injecting a chemical cocktail that generates abnormal growth in the plant itself. They prefer soft new growth, so shoot tips will look distorted, swollen, red-pink and even hairy. The distortion often stops any flowers from forming and can quickly spread all over the plant and from there to any other fuchsias in the neighbourhood.

Fuchsia gall mites are fans of warm weather and can usually be seen from May to September, but will happily overwinter in mild areas on the host plant, hiding within scales and buds.

Multiplying mites

Fuchsia gall mites have a 21-day life cycle and each female can lay 50 eggs at a time, so it does not take long for a problem to form. It is estimated that from one female the number of mites can expand to around 125,000 in just three generations – that's under three months!

CHECK, AND CHECK AGAIN

Be vigilant when buying new fuchsias. Check the plants thoroughly and always buy from reputable suppliers. Avoid cuttings unless you are 100% confident that the parent plants are clean.

Humans make the perfect transport for these microscopic gall mites, stomping around gardens, brushing past infected plants or wielding tools, so take care when around fuchsias. Buy resistant varieties if you can.

WHAT YOU CAN DO

If you come across an infestation, try cutting off as many of the infested tips as you can. Repeat this weekly as necessary and cross your fingers. Remove any seriously infested plants completely, seal them in a plastic bag and leave to die off in the sun. Ask your neighbours to do the same if they have fuchsias, as gall mites can travel on the wind and be carried by bees.

Also, prune fuchsias in autumn severely (down to ground level) and dispose of the old season's growth to prevent the pest from overwintering.

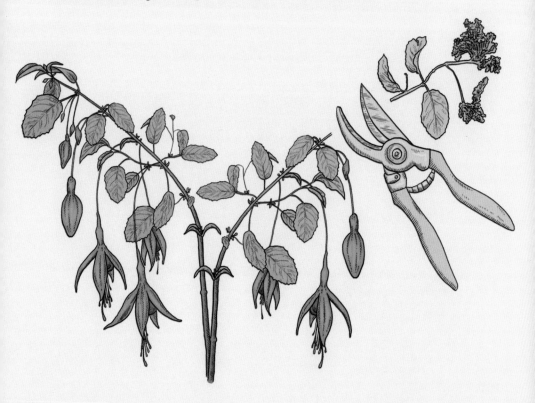

47 Go on night patrol to catch sneaky weevils

✳ quick fix   🪲 adult vine weevils

Vine weevils (*Otiorhynchus sulcatus*), both as adult beetles and juvenile larvae, can attack almost any plant, and particularly favour pot plants. Vigilance is key to preventing them from causing damage.

WILY WEEVILS

Members of the beetle family, adult vine weevils are about 1cm (¹/₂in) long with dull, black bodies, a long snout and jointed antennae (see below). They are slow moving, unable to fly and crawl from plant to plant nibbling on leaves and leaving irregular notches on leaf edges. The damage the adults do is fairly minimal, but during the course of their summer wanderings, they will also manage to lay many hundreds of eggs in the soil around chosen plants.

WAYS TO CATCH WEEVILS

They are most active from late spring to autumn but tend to hide away during the day. This means that hunting for them involves regular night-time forays into the garden with a torch and a container to collect them. Sticky barrier glues on trunks or around pots can help to stop the adults moving around, but these need to be kept free of debris to work successfully. Try a thick layer of grit on the surface of pots likely to be susceptible to vine weevil attack – this may discourage the adults from laying their eggs nearby.

Weevil wish list

These plants are likely to feature on a vine weevil's menu:

- Azalea (*Rhododendron*)
- Begonia (*Begoniaceae*)
- Cyclamen (*Primulaceae*)
- Euonymus (*Euonymus europaeus*)
- Heuchera (*Saxifragaceae*)
- Hostas (*Hosta*)
- Hydrangea (*Hydrangea macrophylla*)
- Impatiens (*Impatiens walleriana*)
- Polyanthus (*Micranthocereus polyanthus*)
- Primula (*Primula*)
- Rhododendrons (*Rhododendron ferrugineum*)
- Strawberry (*Fragaria × ananassa*)
- Indoor plants, especially those put outside in the summer

Euonymus
(*Euonymus europaeus*)

Heuchera
(*Saxifragaceae*)

Azalea
(*Rhododendron*)

Hydrangea
(*Hydrangea macrophylla*)

48 Invite vine weevil larvae's natural enemies

** Fairly simple   vine weevil larvae

Juvenile vine weevils (*Otiorhynchus sulcatus*) at larval stage inflict far more damage than when they are adult beetles. However, there are some tiny natural enemies that you can introduce to hunt them out.

DAMAGE DONE

Vine weevil eggs hatch around August into larvae that fatten up on the roots or tubers of many plants. They can feed all the way through to March in mild conditions or in a greenhouse, although cold weather will slow them down. Once fully grown, they will burrow deeper and pupate into adults that emerge once again ready to lay as the weather warms.

Unless you spot one of these creamy-white, legless, curved grubs while cultivating the soil, the first signs of a vine weevil problem will be a general lack of vigour, as if the plant has not had enough water, followed by wilting and death. Lift the affected plant and you will quickly see that the roots have been completely eaten away.

Creating masses

Vine weevil larvae are plump, C-shaped and just under 1cm (¹/₂in) long. They are creamy-white with a brown head and are, importantly, flightless. All adults are females and can lay around 1,000 eggs each summer – they don't need males to reproduce. Therefore, it doesn't take long for hundreds of larvae to hatch and start destroying the root system of your favourite plants.

THERE IS HOPE

It is sometimes possible to lift smaller plants before too much damage has been done and check for signs of larvae. If larvae are present but have not destroyed all the roots, remove all the old compost and wash the root gently, along with the pot if applicable, before repotting or replanting elsewhere.

NATURAL BORN KILLERS

Fortunately, vine weevils have enemies, specifically pathogenic nematodes (tiny grub-infesting worms), that if used correctly can successfully control them. Biological control packs of millions of these tiny worms can be obtained, mixed with water and then applied to the soil, where they will go to work on any larvae they come across. They usually prefer warm, moist conditions, but each type comes with its own set of specific instructions. Nothing else will be harmed in the process. These are best applied in late summer or early autumn.

Tiny vine weevil larvae
will eat their way through
the roots or tubers of your
favourite plants.

49 Rotate crops to reduce disease and pest infestation

✳✳✳ effort required   ⌣ various minibeasts

Humans have been moving their crops around for centuries. While this is mainly to ensure that fresh soil is free from disease, it can also help to reduce pest infestations such as potato cyst nematode (*Globodera*).

UTILIZE SPACE

In an ideal world, all crops would be best grown in new soil each year, free from pests, diseases and deficiencies. Growing the same plants year after year in the same place just intensifies any pest, disease and nutrient problems. Simple rotation systems have been devised to work in the space we have available.

Crop rotation is simply dividing whatever space you have into sections and changing the plants you put in each year. You don't have to follow a specific system religiously, as long as you have a good memory and can remember where everything should go or enjoy keeping detailed records. Most people prefer to use a system that groups certain kinds of plants together and has a certain 'easy-to-follow' order to it.

← Microscopic beneficial nematodes can be watered onto recently dug gardens to control wireworms.

CROP INVADERS

Soil-dwelling pests such as potato cyst nematodes do not get the chance to build up over time if the plants they are living on or in are moved somewhere else each year. The same is true of a variety of soil-borne diseases too.

The most popular crop-rotation system involves using four plots or sections:

Plot 1 Root crops such as carrot (*Daucus carota*) and parsnip (*Pastinaca sativa*) as well as leeks (*Allium porrum*) and onions (*Allium cepa*).

Plot 2 Members of the pea (*Pisum sativum*) and bean (*Phaseolus*) family.

Plot 3 Cabbages and other brassicas such as sprouts (*Brassica oleracea* var. *gemmifera*) and broccoli (*Brassica oleracea* var. *italica*).

Plot 4 Potatoes (*Solanum tuberosum*), tomatoes (*Solanum lycopersicum*) and everything else.

There are many versions of this four-bed rotation and the system can be expanded depending on what you choose to grow.

Perennial plants such as rhubarb (*Rheum rhabarbarum*) and asparagus (*Asparagus officinalis*) can be slotted in anywhere there is space and moved to a different location after a few years if there is a build-up of pest problems.

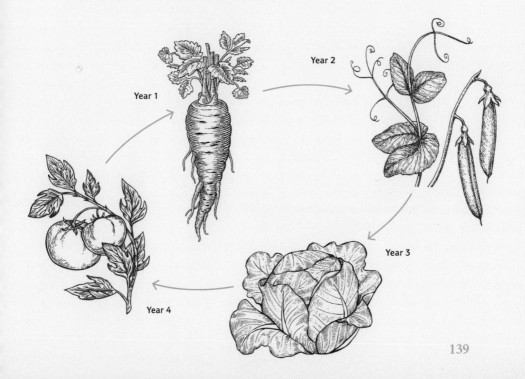

Year 1

Year 2

Year 3

Year 4

139

50 How to handle unsightly anthills

✳✳ fairly simple 🐜 **ants**

Ants are more of an occasional nuisance than a bona fide pest and actually do more good than harm. However, anthills can be very annoying when you want to cut your grass.

OVER THE HILL

If they are not causing a problem, it is better to live and let live, as attempting to remove all the nests in your garden will only make things worse in the long run. Established colonies do a great job of ensuring that few other ants intrude, controlling their own territory and killing any queens that fly in looking for a place to settle.

The main problem tends to be anthills in lawns when excavated soil from the nest appears as little volcanoes. These can cause uneven lumps and make mowing difficult, as the soil is smeared over the grass, especially when wet. The solution is simple. Pick a nice dry day and shovel up as much of the loose soil as you can before brushing the rest across the lawn. Or rake the anthill flat and spread the soil in a fine layer. Repeat whenever you can before mowing.

ANT SHEPHERDS

Ants crawling up and down the trunks and branches of plants are a common sight. They are actually harvesting the sugary honeydew excreted by aphids higher up on the plant. In return for all that sweetness, they look after the aphids, protecting them from predators. You should worry about the aphids, not the ants.

INHABIT-ANTS

Lasius niger – black garden ants prefer sandy dry soil.

Lasius flavus – yellow meadow ants prefer grass that's not cut much; they can make large anthills.

Myrmica rubra – common red ants (the stinging ones!) prefer the underside of large stones or paving slabs, but will nest in soil or grass.

Anti-ant aromas

It has been claimed that ants dislike strong-smelling plants, so try orange (*Citrus × sinensis*) peel or mint (*Mentha*) sprigs around any problem nests. With luck they will march on.

Index

Acknowledgements

IMAGE CREDITS

t = top, m = middle, b = bottom, l = left, r = right, tl = top left, tr = top right, bl = bottom left, br = bottom right

Alamy: Alan Mather 39 b; garfotos 63 l; fotolincs 69; Parmorama 81 t; Roland Smithies 81 m; Florapix 99 tl; christopher miles 124 l; Dave Bevan 137

Getty Images: mikroman6 134

iStock: peplow 29 r; ilbusca 106

Shutterstock: Morphart Creation 1; 13 b, 68, 94 l, 95 l, 103 t, 112, 115 b, 135 l; Oleksii Borodachov 13 t; Andrew Pustiakin 26 l; Furiarossa 26 r; Alexander_P 27 b, 139 l; Gagarina Vasilisa 28; Badon Hill Studio 29 l; Nicolette_Wollentin 39 t; Bodor Tivadar 58, 109, 130; inimma 63 r; Steve Cymro 70; mamita 71; D Maistruk 72; Rudmer Zwerver 78; All-stock-photos 81 b; krolya25 83 t; RealityImages 83 m; Thijs de Graaf 83 b; David Orcea 87 tl; Wagner Campelo 87 tr; Anna_Huchak 87 bl; TristanBM 87 br; CTatiana 88 tl; Scisetti Alfio 88 tr; Shebeko 88 bl; Tritippayanipha Thani 88 br; Tibesty 89 tl; Tatyana Mi 89 tr; Tanya May 89 bl; Zorica Popovic 89 br; Katunina 90; flagman_1 94 t; ahmydaria 94 m; MityaChernov 94 b; Liliana Akstein 95 t; Maria Wan 95 m; Starover Sibiriak 95 b; Fabian Junge 99 tr; Animaflora PicsStock 99 bl; olegpodi 99 br; KUCO 104; M Rose 107 t; Svitlyk 107 b; Orest lyzhechka 111; Tomasz Klejdysz 114; andrey oleynik 115 t; Peter Wey 120; taviphoto 124 r; Heiti Paves 128; photolike 135 t; AliScha 135 m; imageportal 135 b; Helga Fluey 138; Ianrward 139 t; NataLima 139 r; ZVERKOVA 139 b

AUTHOR ACKNOWLEDGEMENTS

I would like to thank Claire Boothby from the Bat Conservation Trust for all her advice and help with information on bats, birds and any general wildlife queries. Also, Ali Marshall, head gardener at Torre Abbey Gardens, for her tips and suggestions on organic methods of controlling garden pests and for encouraging wildlife into the garden. I would also like to say a massive thank you to Guy Barter and the RHS team for their advice, and also to James, Caroline, Lindsey and Rica from Quarto for all their hard work, and Ian for his beautiful illustrations.